The ESSEN
AUTOMATIC CONTROL SYSTEMS / ROBOTICS I

**Staff of Research and Education Association,
Dr. M. Fogiel, Director**

> This book covers the usual course outline of Automatic Control Systems/ Robotics I. For more advanced topics, see *"THE ESSENTIALS OF AUTOMATIC CONTROL SYSTEMS/ROBOTICS II"*.

 Research and Education Association
61 Ethel Road West
Piscataway, New Jersey 08854

THE ESSENTIALS OF AUTOMATIC CONTROL SYSTEMS/ROBOTICS I

Copyright © 1987 by Research and Education Association. All rights reserved. No part of this book may be reproduced in any form without permission of the publishers.

Printed in the United States of America

Library of Congress Catalog Card Number 87-61817

International Standard Book Number 0-87891-571-0

WHAT "THE ESSENTIALS" WILL DO FOR YOU

This book is a review and study guide. It is comprehensive and it is concise.

It helps in preparing for exams, in doing homework, and remains a handy reference source at all times.

It condenses the vast amount of detail characteristic of the subject matter and summarizes the **essentials** of the field.

It will thus save hours of study and preparation time.

The book provides quick access to the important facts, principles, theorems, concepts, and equations of the field.

Materials needed for exams, can be reviewed in summary form — eliminating the need to read and re-read many pages of textbook and class notes. The summaries will even tend to bring detail to mind that had been previously read or noted.

This "ESSENTIALS" book has been carefully prepared by educators and professionals and was subsequently reviewed by another group of editors to assure accuracy and maximum usefulness.

Dr. Max Fogiel
Program Director

CONTENTS

Chapter No.		Page No.
1	**SYSTEM MODELING: MATHEMATICAL APPROACH**	1
1.1	Electric Circuits and Components	1
1.2	Mechanical Translation Systems	3
1.3	Mechanical and Electrical Analogs	5
1.4	Mechanical Rotational Systems	6
1.5	Thermal Systems	7
1.6	Positive-Displacement Rotational Hydraulic Transmission	9
1.7	D-C and A-C Servomotor	10
1.8	Lagrange's Equation	14
2	**SOLUTIONS OF DIFFERENTIAL EQUATIONS: SYSTEM'S RESPONSE**	16
2.1	Standardized Inputs	16
2.2	Steady State Response	16
2.3	Transient Response	20
2.4	First and Second-Order System	24
2.5	Time-Response Specifications	28

3 APPLICATIONS OF LAPLACE TRANSFORM 29

3.1	Definition of Laplace Transform	29
3.2	Application of Laplace Transform to Differential Equations	32
3.3	Inverse Transform	33
3.4	Frequency Response from the Pole-Zero Diagram	38
3.5	Routh's Stability Criterion	39
3.6	Impulse Function: Laplace-Transform and its Response	41

4 MATRIX ALGEBRA AND Z-TRANSFORM 44

4.1	Fundamentals of Matrix Algebra	44
4.2	Z-Transforms	47

5 SYSTEM'S REPRESENTATION: BLOCK DIAGRAM, TRANSFER FUNCTIONS, AND SIGNAL FLOW GRAPHS 51

5.1	Block Diagram and Transfer Function	51
5.2	Transfer Functions of the Compensating Networks	53
5.3	Signal Flow Graphs	55

6 SERVO CHARACTERISTICS: TIME-DOMAIN ANALYSIS 58

6.1	Time-Domain Analysis Using Typical Test Signals	58
6.2	Types of Feedback Systems	59

| 6.3 | General Approach to Evaluation of Error | 60 |
| 6.4 | Analysis of Systems: Unity Feedback | 63 |

7 ROOT LOCUS 67

7.1	Roots of Characteristic Equations	67
7.2	Important Properties of the Root Loci	69
7.3	Frequency Response	78

8 SPECIAL POLE-ZERO TOPICS: DOMINANT POLES AND THE PARTITION METHOD 80

8.1	Transient Response: Dominant Complex Poles	80
8.2	Pole-Zero Diagram and Frequency and Time Response	85
8.3	Factoring of Polynomials Using Root-Locus	87

CHAPTER 1

SYSTEM MODELING: MATHEMATICAL APPROACH

1.1 ELECTRIC CIRCUITS AND COMPONENTS

The equations for an electric circuit obey Kirchoff's laws which can be stated as follows:

A) Σ potential differences around a closed circuit = 0

B) Σ currents at a junction or node = 0

Voltage drops across three basic electrical elements:

A) $v_R = Ri$

B) $v_L = L \frac{di}{dt} = LDi$

C) $v_C = \frac{q}{c} = \frac{1}{c} \int_0^t i \, d\tau + \frac{Q_o}{C} = \frac{1}{c} \cdot \frac{i}{D}$

where operator D is defined as the following:

$$DX = \frac{dx}{dt}$$

and $\quad \dfrac{X}{D} \equiv \displaystyle\int_{-\infty}^{t} x\, d\tau$

SERIES R-L CIRCUIT

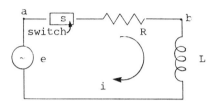

Fig. Simple R-L circuit

$$Ri + L\dfrac{di}{dt} = e = \dfrac{R}{L} \cdot \dfrac{v_L}{D} + v_L$$

(because $i_{inductor} = \dfrac{1}{LD} v_L$)

$$\dfrac{v_b - v_a}{R} + \dfrac{1}{L} \cdot \dfrac{v_b}{D} = 0 \quad \ldots \text{ Kirchoff's second law}$$

$$\left(\dfrac{1}{R} + \dfrac{1}{LD}\right) v_b - \dfrac{v_a}{R} = 0$$

SERIES R-L-C CIRCUIT

$$v_R + \dfrac{L}{R} Dv_R + \dfrac{1}{RC} \cdot \dfrac{v_R}{D} = e$$

MULTILOOP ELECTRIC CIRCUITS

Loop method:

A loop current is drawn in each closed loop; then Kirchoff's voltage equation is written for each loop. These equations are solved simultaneously to obtain output (voltage) in terms of input (voltage) and the circuit parameters.

Node Method:

The rules for writing the node equations:

A) The number of equations required is equal to the number of unknown node voltages.

B) An equation is written for each node.

C) The equation includes the following terms: the node voltage multiplied by the sum of all the admittances that are connected to this node, and the node voltage of the other end of each brand multiplied by the admittance connected between the two nodes.

1.2 MECHANICAL TRANSLATION SYSTEMS

The mechanical translation system is characterized by mass, elastance and damping.

Representation of the basic elements:

A) Mass

    a) It is the inertial element.

    b) Reaction force f_M = M x acceleration = Ma

$$= M \frac{dv}{dt} = M \frac{d^2 s}{dt^2} ,$$

    c) Network representation:

a has the motion of the mass

b has the motion of the reference

f_M is a function of time

B) Elastance (or stiffness, k):

    a) Representation:

b) Reaction force $f_k = k(x_c - x_d)$; x_c is the position of c and x_d is the position of d.

C) Damping (viscous friction, B):

   a) Representation:

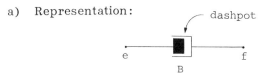

   b) The reaction damping force $f_B = B(v_e - v_f)$
   $$= B(Dx_e - Dx_f)$$

SIMPLE TRANSLATION SYSTEM

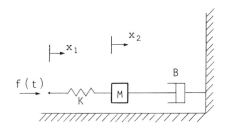

$$f = f_k = k(x_a - x_b)$$
$$f_k = f_M + f_B = MD^2x_2 + BDx_2$$

These can be solved for displacements x_1 and x_2 and Dx_1 and Dx_2.

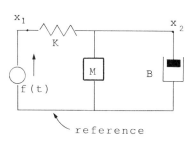

Fig: Mechanical Network

The system is initially at rest. To draw the mechanical method, x_1 and x_2 and the reference are located.

MULTI-ELEMENT SYSTEM

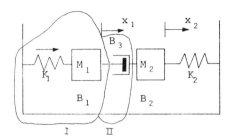

The system equations must be written in terms of two displacements x_1 and x_2. These are the nodes in the equivalent mechanical network:

$$(M_1 D^2 + B_1 D + B_3 D + K_1)x_1 - (B_3 D)x_2 = f \ldots$$

because the forces at node x_1 must add to zero. At node x_2, the equations are written by observing the above pattern:

$$(M_2 D^2 + B_2 D + K_2)x_2 - (B_3 D)x_1 = 0$$

Using these, an equivalent mechanical network can be drawn using the following table of electrical and mechanical analogies.

1.3 MECHANICAL AND ELECTRICAL ANALOGS

Mechanical Element	Electrical Element
M – mass	C – capacitance
f – force	i – current
$v = \dfrac{dv}{dt}$ – velocity	e or v – voltage
B – damping coefficient	$G = \dfrac{1}{R}$ – conductance
K – stiffness coefficient	$\dfrac{1}{L}$ = reciprocal inductance

An equivalent electrical network is drawn using the table of electrical and mechanical analogs.

These two networks have the same mathematical forms.

1.4 MECHANICAL ROTATIONAL SYSTEMS

Network elements of mechanical rotational systems:

```
a ———[ J ]——— b          J: Moment of inertia

c ——/\/\/\——— d
        K

e ———[==]———— f
        B
```

A) The reaction torque = $T_J = J\alpha = JD\omega = JD^2\theta$

   where θ is the angular displacement.

B) Reaction spring torque = $T_k = K(\theta_c - \theta_d)$

   where θ_c and θ_d are the positions of the two ends of a spring ($\theta_c - \theta_d$) = the angle of twist.

C) Damping torque $T_B = B(\omega_e - \omega_f)$...

   where $\omega_e - \omega_f$ = relative angular velocity of the ends of the dashpot.

SIMPLE ROTATIONAL SYSTEM

The governing equation is

$$JD^2\theta + BD\theta + K\theta = T(t)....$$

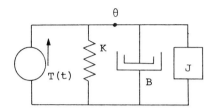

Only one equation is necessary because the system has only one node. The actual system consists of a shaft with fins of a moment of inertia, J, which is immersed in oil. The fluid has a damping factor of B.

1.5 THERMAL SYSTEMS

Only a few thermal systems are represented by linear differential equations. The basic requirement is that the temperature of the body should be assumed to be uniform.

NETWORK ELEMENTS

C: Thermal capacitance
R: Thermal resistance

A) h = Heat stored = $\frac{q}{D} C(T_2 - T_1)$ due to change in temperature $(T_2 - T_1)$

B) $q = CD(T_2 - T_1)$ in terms of rate of heat flow.

C) q = Rate of heat flow = $\frac{T_3 - T_4}{R}$ where T_3 and T_4 are two boundary temperatures.

Temperature is analogous to potential.

SIMPLE THERMAL SYSTEM: MERCURY THERMOMETER

A) Network representation

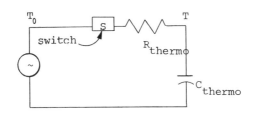

Fig: Simple Network

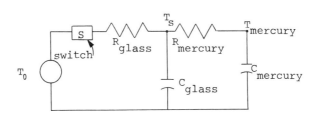

Fig: Exact Network

B) q = flow of heat = $(T_0 - T)/R$ where T_0 is the temperature of the bath and T is the temperature before immersing into the bath.

C) h = Heat entering the thermometer = $C(T - T_1)$

D) $RCDT + T = T_0$ is the governing equation.

E) More exact analysis:

   T_s = The temperature at the inner surface between the glass and the mercury

   For node s: $\left(\dfrac{1}{R_g} + \dfrac{1}{R_m} + C_g D \right) T_s - \dfrac{T_m}{R_m} = \dfrac{T_o}{R_g}$

   For node m: $\dfrac{-T_s}{R_m} + \left(\dfrac{1}{R_m} + DC_m \right) T_m = 0$

$$\left[C_g C_m D^2 + \left(\frac{C_g}{R_m} + \frac{C_m}{R_g} + \frac{C_m}{R_m} \right) D + \frac{1}{R_g R_m} \right] T_m$$

$$= \frac{T_o}{R_g R_m} \quad \text{the governing equation.}$$

The latter equation is from the form

$$(A_2 D^2 + A_1 D + A_o) T_m = k T_o.$$

1.6 POSITIVE-DISPLACEMENT ROTATIONAL HYDRAULIC TRANSMISSION

The hydraulic transmission is used when a large torque is required. It contains a variable displacement pump driven at a constant speed. It is assumed that the transmission is linear over a limited range.

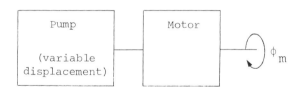

A) $q_p = q_m + q_1 + q_c$ Since the fluid flow rate from the pump must equal the sum of the flow rates.

$q_p = x d_p \cdot \omega_p$

$q_m = d_m \cdot \omega_m$, $q_1 = LP_L$ and $q_c = Dv = \dfrac{vDP_L}{K_B}$

B) Torque at the motor shaft $= T = n_T \cdot d_m \cdot P_L$

$$= CP_L$$

where P_L is the load-induced pressure-drop across motor.

DIFFERENT CASES

A) Inertia load

$$T = J \cdot D^2\phi_m = CP_L$$

B) Inertia load coupled through a spring

$$T = K(\phi_m - \phi_L) = JD^2\phi_L = CP_L$$

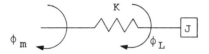

C) In case of inertia load, value of P_L obtained from equation $q_p = q_m + q_1 + q_c$ is substituted into $T = CP_L$. The resulting equation can be solved for ϕ_m in terms of x. The same procedure is adopted in case of a spring-coupled inertia load.

1.7 D-C AND A-C SERVOMOTOR

D-C SERVOMOTOR

$T(t)$ = The torque = $Ka\phi\, i_m \ldots i_m$ = armature current, ϕ = the flux, K_a = constant of proportionality

Modes of Operation

A) An adjustable voltage is applied to the armature while the field current is held constant.

B) An adjustable voltage is applied to the field while the armature current is kept constant.

Armature Control

A constant field current is obtained by exciting the field from a fixed Dc source.

$T(t) = k_T i_m$ where ϕ is a constant (since field current is a constant)

Back emf = $e_m = k_i \phi \omega_m = k_b \omega_m = kb \cdot D\theta_m$ ... ω_m = speed

Armature-controlled d-c motor:

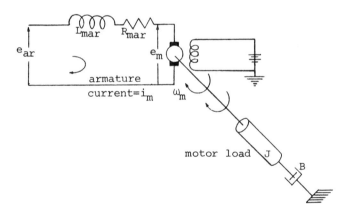

$e_{ar} = L_m(L_{mar} + R_{mar}) + e_m$

$JD\omega_m + B\omega_m = T(t)$ ... Torque equation with motor load.

$$\frac{L_{mar} J}{k_T} D^3\theta_m + \frac{L_{mar} B + R_{mar} J}{k_T} D^2\theta_m + \frac{R_{mar} B + K_b K_T}{k_T} D\theta_m = e_{ar}$$

The system equation.

Field Control

In this case the armature current i_{mar} is constant so that T(t) is proportional only to the flux ϕ:

$$T(t) = K_3 \cdot \phi \cdot i_{mar} = K_3 K_2 \cdot i_{mar} \cdot i_{field} = k_f \cdot i_{field}$$

where i_{mar} is constant.

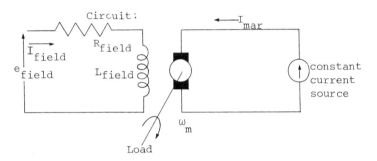

Circuit.

$$e_{field} = L_{field} D_{field} + R_{field} \cdot i_{field}$$

$$(K_f \cdot D + R_{field})(JD + B)\omega_m = k_f \cdot e_{field}$$

The system equation.

A-C SERVOMOTOR

This may be considered to be a two-phase induction motor having two field coils positioned 90 electrical degrees apart.

In a two-phase induction motor, the speed which is a little below the synchronous speed is constant; however when the unit is used as a servomotor, the speed is proportional to the input voltage.

The two-phase induction motor shown below can be used as a servomotor by applying an ac voltage e to one of the windings. Thus e is fixed and when e_{c_0} is varied, the torque and speed are a function of this voltage.

Because these curves are non-linear, we must approximate them by straight lines in order to obtain linear differential equations.

Energy functions for electric circuits based on the loop or mesh analysis

Table 1.2

Element	Kinetic energy T	Potential energy V	Dissipation function D	Forcing function Q
Voltage source, e	—	—	—	e
Inductance, L	$\frac{1}{2}Li^2 = \frac{1}{2}L\dot{q}^2$ where q is the charge	—	—	—
Capacitance, C	—	$\dfrac{(\int i\,dt)^2}{2C} = \dfrac{q^2}{2C}$	—	—

CHAPTER 2

SOLUTIONS OF DIFFERENTIAL EQUATIONS: SYSTEM'S RESPONSE

2.1 STANDARDIZED INPUTS

Sinusoidal function: $r(t) = \cos\omega t$

Power-series function: $r = a_0 + a_1 t + a_2 t^2 + \ldots$

Unit step function: $r = u(t)$

Unit ramp function: $r = tu(t)$

Unit parabolic function: $r = t^2 u(t)$

Unit impulse function: $r = \delta(t)$

2.2 STEADY STATE RESPONSE

SINUSOIDAL INPUT

Input: $r(t) = A\cos(\omega t + \alpha)$

This is generally the form of input.

$$= \text{real part of } (Ae^{j(\omega t + \alpha)}) = \text{Re}(Ae^{j(\omega t + \alpha)})$$

$$= \text{Re}(Ae^{j\alpha} e^{j\omega t}) = \text{Re}(\mathbf{A}e^{j\omega t})$$

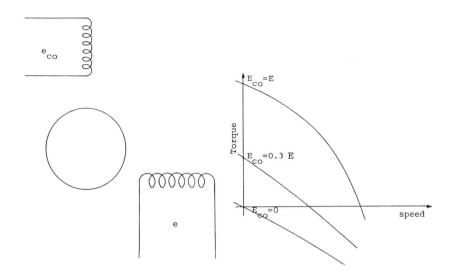

From these curves, it is noted that the generated torque is a function of the speed ω and voltage e_{c_0}.

The torque equation is given by

$$\boxed{e_{c_0} \frac{\partial T}{\partial e_{c_0}} + \frac{\partial T}{\partial \omega} \omega = T(e_{c_0}, \omega)} \qquad (a)$$

Because of the straight line approximation we are using, it is justifiable to let

$$\frac{\partial T}{\partial e_{c_0}} = K_{c_0} \quad \text{and} \quad \frac{\partial T}{\partial \omega} = K_\omega$$

Suppose we assume a load consisting of inertia and damping, then

$$T_L = JD\omega + B\omega \qquad (b)$$

However, the generated torque must be equal to the load torque. Therefore, equating equations (a) and (b) we get

$$K_{c_0} e_{c_0} + K\omega = JD\omega + B\omega$$

$$\boxed{JD\omega + (B - K)\omega = K_{c_0} e_{c_0}}$$

1.8 LAGRANGE'S EQUATION

$$\frac{d}{dt}\left(\frac{\partial T}{\partial \dot{q}_n}\right) - \frac{\partial T}{\partial q_n} + \frac{\partial D}{\partial \dot{q}_n} + \frac{\partial V}{\partial q_n} = Q_n$$

where $n = 1, 2, 3, \ldots$ are the independent coordinates or degrees of freedom which exist in the system and

- T = total kinetic energy of system
- D = dissipation function of system
- V = total potential energy of system
- Q_n = generalized applied force at the coordinate n
- q_n = generalized coordinate
- $\dot{q}_n = dq_n/dt$ (generalized velocity)

Energy functions for translational mechanical elements
Table 1.1

Element	Kinetic energy T	Potential energy V	Dissipation factor D	Forcing function Q
Force, f	—	—	—	f (force)
Mass, M	$\tfrac{1}{2}Mv^2 = \tfrac{1}{2}M\dot{x}^2$	—	—	—
Spring K, x_1, x_2	—	$\dfrac{K}{2}\left[\int (v_1 - v_2)dt\right]^2$ $= \tfrac{1}{2}K(x_1 - x_2)^2$ where x is the displacement	—	—
Damping, B, v_1, v_2	—	—	$\tfrac{1}{2}B(v_1 - v_2)^2$ $= \tfrac{1}{2}K(\dot{x}_1 - \dot{x}_2)^2$	—

using Euler's identity.

Form of the equation to be solved:

$$A_m D^m C + A_{m-1} D^{m-1} C \ldots + A_{-n} D^{-n} C = r$$

$\bar{R} = Re^{j\alpha}$ phasor representation of the input.

$\bar{C}$ = Phasor representation of the output

$$= \frac{\bar{R}}{A_m (j\omega)^m + A_{m-1}(j\omega)^{m-1} \ldots + A_{-n}(j\omega)^{-n}}$$

Time response = $c(t) = |\bar{C}| \cos(\omega t + \phi)$

Steady-State Sinusoidal Response of Series RLC Circuit:

$$LDi + Ri + \frac{1}{CD} i = e$$

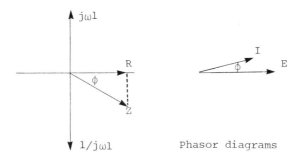

Phasor diagrams

The equation obtained after application of Kirchoff's voltage law.

$$e = Re[\sqrt{2}\,\bar{E}(j\omega)e^{j\omega t}]$$

The input to the RLC circuit.

$$\bar{I}(j\omega) = \text{Phasor current} = \frac{\bar{E}(j\omega)}{j\omega L + R + \frac{1}{j\omega c}}$$

$$\bar{Z}(j\omega) = j\omega L + R + \frac{1}{j\omega c}$$

Frequency transfer function:

$$\overline{G}(j\omega) = \frac{\overline{V}_R(j\omega)}{\overline{E}(j\omega)} = \frac{R}{j\omega L + R + \frac{1}{j\omega c}}$$

$\overline{V}_R(j\omega)$ is the voltage (output) across R in phasor form.

POWER SERIES INPUT

Consider the general differential equation:

$$A_m D^m C + \ldots + A_0 C + A_{-1} D^{-1} C + \ldots + A_n D^{-n} C = r$$

where $r(t) = a_0 + a_1 t + a_2 t^2 + \ldots + a_k t^k$.

To find the particular (steady state) solution of the response c(t), we assume a solution of the form

$$(H) = b_0 + b_1 t + b_2 t^2 + \ldots + b_v t^v$$

Substituting in the differential equation and equating coefficients on both sides will enable us to find the constants $b_0, b_1, \ldots$.

Note: 1) k is the highest power of t on the right side. Therefore t^k must also appear on the left side of the equation.

     2) The highest power of t on the left side will result from the lowest order derivative. Let x be the order of the lowest derivative.

$$v = k + x \qquad v \geq 0$$

Special Cases of Power Series Input

    (i) Step-Function Input:

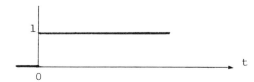

A) Step-function input:

Voltage equation: $A_2 D^2 \omega_m + A_1 D \omega_m + A_0 \omega_m = u(t)$
where u(t) is the input.

The response is of the form $\omega_m = b_o$ because, for step f_n, the highest exponent of t is $k = 0$ for u(t).

$$D \omega_m = 0 = D^2 \omega_m$$

$$\boxed{b_o = \frac{1}{A_o}}$$

B) Ramp-function:

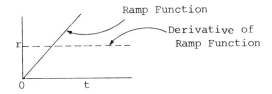

Assume we have a system whose equation is

$$C \cdot D\theta + A = m$$

By integrating, the response is

$$C \cdot \theta + \left(ns + \frac{1}{R}\right) D^{-1}\theta = q$$

One form of the response is

$$\boxed{\theta = b_o}$$

because the highest exponent of t in the input n = t is k = 1.

After integrating,

$$D^{-1}\theta = b_o t + C_o$$

$$b_o = \frac{1}{A}, \quad \theta = \frac{1}{A} \quad \text{at a steady state.}$$

C) Parabolic-function input:

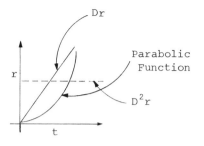

$$r = t^2 u(t) \quad D^2 r = 2u(t)$$

$$Dr = 2t \cdot u(t)$$

Example: Given a system whose equation is

$$AD^2 x_2 + BDx_2 + Kx_2 = Kx_1$$

k = 2 because the highest exponent of t in the input is 2.

Order of the lowest derivation in the system equation

$$x = 0$$

The form of the steady-state response is

$$\boxed{x_2(t) = b_0 + b_1 t + b_2 t^2}$$

$$Dx_2 = b_1 + 2b_2 t \quad \text{and} \quad D^2 x_2 = 2b_2$$

Steady-state solution:

$$x_2(t) = \frac{-2A}{K} + \frac{2B^2}{K^2} - \frac{2Bt}{K} + t^2.$$

2.3 TRANSIENT RESPONSE

Transient response of a differential equation:

REAL ROOTS

Steps:

1) Write a homogeneous equation by equating the given

differential equation to zero.

Differential equation:
$$b_m D^m c + b_{m-1} D^{m-1} c + \ldots + b_o D^o c + \ldots + b_{-n} D^{-n} c = r$$

Homogeneous equation:
$$b_m D^m c_t + b_{m-1} D^{m-1} c_t + \ldots + b_{-n} D^{-n} c_t = 0$$

2) Solution of the homogeneous equation gives the general expression for the transient response.
Assume a solution $C_t = e^{kt}$.

3) Substituting $C_t = e^{kt}$ in the equation results in the characteristic equation:
$$b_m k^m + b_{m-1} k^{m-1} + \ldots + b_o + \ldots + b_{-n} k^{-n} = 0$$

Its roots are $k_1, k_2 \ldots$

4) So if there are no multiple roots the transient response is $C_t = A_1 e^{k_1 t} + A_2 e^{k_2 t} \ldots$

Short-cut method:

We can obtain the characteristic equation by substituting into the homogeneous equation: k for DC_t, k^2 for $D^2 C_t$ etc. For the general equation we are using, our characteristic equation will consist of m + n constants. We must consequently have m + n initial conditions in order to set up m + n equations which will enable us to determine the transient response.

COMPLEX ROOTS

When the roots of the characteristic equation are complex, the above method cannot be used; instead, the response takes the form

$$Ae^{\sigma t} \sin(\omega_d t + \phi) \quad \text{(For 2 complex roots)}$$

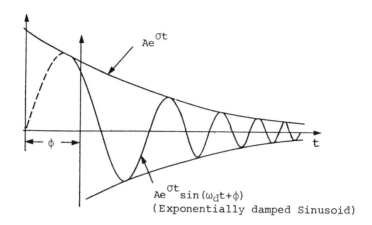
(Exponentially damped Sinusoid)

Damping ratio and undamped national frequency:

$$\text{Damping ratio } \zeta = \frac{\text{Actual damping}}{\text{Critical damping}}$$

If the characteristic equation has a pair of complex conjugate roots, the form of the quadratic factor is $b_2 m^2 + b_1 m + b_0$.

b_1 represents the effective damping constant of the system.

$$\zeta = \frac{\text{Actual damping}}{\text{Critical damping}} = \frac{b_1}{b_1^*} = \frac{b_1}{2\sqrt{b_2 b_0}}$$

If $\zeta < 1$, the response is said to be underdamped.
If $\zeta > 1$, the response is said to be overdamped.
If $\zeta = 0$,

A) $\omega_m = \sqrt{\dfrac{b_0}{b_2}}$ = undamped natural frequency

B) the zero damping constant means that the transient response is a sine wave of constant amplitude.

We can rewrite the above equation $b_2 m^2 + b_1 m + b_0$ as follows:

$$\frac{b_2}{b_0} m^2 + \frac{b_1}{b_0} m + 1 = \frac{1}{\omega_n^2} m^2 + \frac{2\zeta m}{\omega_n} + 1$$

where

$$\omega_m = \sqrt{\frac{b_0}{b_2}}$$

→ $m^2 + 2\zeta\omega_n m + \omega_n^2$

This is the general form of the characteristic equation whose roots are:

$$m_{1,2} = -\zeta\omega_n \pm j\omega_n \sqrt{1-\zeta^2}$$

Transient response for underdamped case:

$$Ae^{-\zeta\omega_n t} \sin(\omega_n \sqrt{1-\zeta^2}\, t + \phi)$$

ω_d – the damped frequency of oscillation

$$\omega_d = \omega_n \sqrt{1-\zeta^2}$$

Time constant:

The transient terms have the form Ae^{kt}. The value of time that makes the exponent of e equal to -1 is called the time constant T.

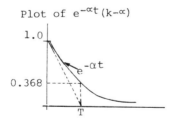

Plot of $e^{-\alpha t}$ $(k-\alpha)$

In case of the damped sinusoid

$$T = \frac{1}{|\sigma|}, \quad \text{for } k = \sigma + \omega_d$$

$$T = \frac{1}{\zeta}\omega_n$$

where the larger the $\zeta\omega_n$, the greater the rate of decay of the transient.

23

2.4 FIRST AND SECOND-ORDER SYSTEM

FIRST-ORDER SYSTEM

An example:

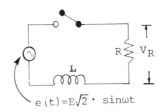

$e(t) = E\sqrt{2} \cdot \sin\omega t$

Applying KVL to series RL circuit,

$$e = \frac{L}{R} Dv_R + v_R$$

The characteristic equation is therefore

$$\frac{L}{R} x + 1 = 0 \quad \text{or,} \quad x = -\frac{R}{L}$$

So the transient solution is

$$v_{R,t} = Ae^{-(R/L)t}$$

The steady-state solution in phasor form is,

$$\overline{V}_{R,SS}(j\omega) = \frac{E(j\omega)}{1 + \left(\frac{L}{R}\right) j\omega} = \frac{E(j\omega)}{\left[1 + \left(\frac{\omega L}{R}\right)^2\right]} \; \underline{/-\tan^{-1} \frac{\omega L}{R}}$$

Therefore, the steady-state voltage in time domain is

$$v_{R,SS} = \frac{E\sqrt{2}}{\left[1 + \left(\frac{\omega L}{R}\right)^2\right]^{\frac{1}{2}}} \sin\left(\omega t - \tan^{-1} \frac{\omega L}{R}\right)$$

The complete solution is:

$$\boxed{V_R = \frac{E\sqrt{2}}{1 + \left(\frac{\omega}{L}\right)^2} \sin\left(\omega t - \tan^{-1}\frac{\omega L}{R}\right) + Ae^{-\frac{R}{L}t}}$$

|← steady-state solution →| |← transient →|
                                  solution

Evaluation of A using initial conditions:

initial conditions: $V_R = 0$ at $t = 0$

$$A = \frac{\omega \cdot R \cdot L \cdot E\sqrt{2}}{R^2 + (\omega L)^2}$$

obtained after putting $V_R = 0$ and $t = 0$ in the complete solution.

SECOND-ORDER SYSTEM

An example

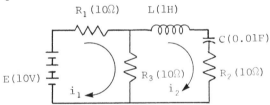

$$E = i_2\left(R_1 + \frac{R_1 R_2}{R_3} + \frac{R_1}{R_3}LD + \frac{R_1}{CDR_3} + R_2 + LD + \frac{1}{CD}\right)$$

$$10 = \left(2D + 40 + \frac{200}{D}\right)i_2 \quad \text{after application of KVL.}$$

The steady-state output is: $i_{2,ss} = 0$...since the branch contains a capacitor.

The characteristic equation: $m + 20 + \frac{100}{m} = 0$.

Its roots are $m_{1,2} = -10$...so the circuit is critically damped.

The output current:

$$i_2(t) = A_1 e^{-10t} + A_2 t e^{-10t}$$

$$Di_2(t) = -10A_1 e^{-10t} + A_2(1 - 10t)e^{-10t}$$

Initial conditions: $i_1(0^-) = i_2(0^-) = 0$

$$i_2(0^+) = i_2(0^-) = 0$$

because i_2 can't change instantly in the inductor.

A) $Di_2(t) = 10i_1(t) - 25i_2(t) - vi(t)$

B) $i_1(0^+) = 0.5$ because $i_2(0^+) = 0$

C) $v_c(0^-) = v_c(0^+)$ = steady-state value of = 10 capacitor voltage for $t < 0$

Hence, $Di_2(t) = -5$

$A_1 = 0$, $A_2 = -5$, so $i_2(t) = -5t\, e^{-10t}$

Second-order transients:

Simple second-order equation:

$$\boxed{\frac{D^2 c}{\omega_n^2} + \frac{2\zeta}{\omega_n} Dc + c = r}$$

because there are no derivatives of r on the right-hand side.

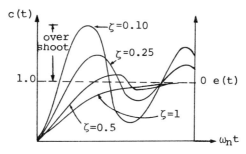

Representation of the Transient Response of a Second Order System

$$c(t) = 1 - \frac{e^{-\zeta\omega_n t}}{\sqrt{1-\zeta^2}} \sin(\omega_n \sqrt{1-\zeta^2}\, t + \cos^{-1}\zeta)$$

This is the response to a unit step with initial conditions set to zero.

$e(t)$ - error in the system

$e = r - c$, where $r = u(t)$

Representation of the transient response of a second-order system.

Conclusion: The amount of overshoot (i.e. beyond 1.0) depends on the damping ratio ζ.

For the overdamped ($\zeta > 1$) and critically damped case ($\zeta = 1$), there is no overshoot.

For underdamped ($\zeta < 1$), the system oscillates around the steady state value before it settles down at the steady state.

$$t_p = \pi/\omega_n \sqrt{1 - \zeta^2} = \text{The time at which the peak overshoot occurs.}$$

$$c_p = 1 + e^{-\zeta\pi/\sqrt{1 - \zeta^2}} \quad \text{The value of the peak overshoot.}$$

$$M_o = \frac{c_p - c_{ss}}{c_{ss}} \quad \text{per unit overshoot.}$$

Peak overshoot v/s ζ and Frequency of Oscillation v/s Damping ratio

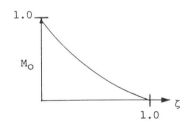
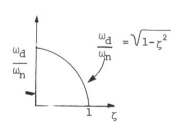

Error equation:

$$e = r - c = \frac{e^{-\zeta\omega_n t}}{\sqrt{1 - \zeta^2}} \sin(\omega_n \sqrt{1 - \zeta^2} \cdot t + \cos^{-1}\zeta)$$

RESPONSE CHARACTERISTICS

One of the characteristics of the system's response is the setting time. Setting time is the time required for the envelope of the transient to die out or to reduce to an insignificantly small value. This time is a function of a number of time constants involved in the transfer function of the system, and can be presented mathematically as follows:

$$T_s = \frac{\text{no. of time constants in the system's transfer function}}{(\text{undamped natural frequency})(\text{the damping constant})}$$

2.5 TIME-RESPONSE SPECIFICATIONS

The following terms are used as specifications to evaluate the performance of a system. These are:

A) Maximum overshoot (Cp): the magnitude of the first overshoot as shown.

B) t_p (Time to reach the maximum overshoot): the time required to reach the first overshoot.

C) Duplicating time (t_o): the time when the response of the system has the final value for the first time.

D) Settling time (t_s): the time taken by the control system to reach the final specified value and thereafter remain within a specified tolerance.

E) Frequency of oscillation (ω_d): the frequency of oscillation of the tansient response.

CHAPTER 3

APPLICATIONS OF LAPLACE TRANSFORM

3.1 DEFINITION OF LAPLACE TRANSFORM

$$L[f(t)] = \int_0^\infty f(t)e^{-st}dt = F(s)$$

provided f(t) is piecewise continuous over every finite interval and is of exponential order.

Laplace Transform of Some Simple Functions:

$L[\text{step function } u(t)] = \frac{1}{s}$ ... if $\sigma > 0$

$L[e^{-\alpha t}] = \frac{1}{s+\alpha}$ .... if $\sigma > -\alpha$

$L[\cos\omega t] = \frac{s}{s^2 + \omega^2}$ ...... if $\sigma > 0$

$L[\text{Ramp function } \dot{f}(t) = t] = \frac{1}{s^2}$ ... if $\sigma > 0$

where s is a complex variable and has the form $\sigma + j\omega$.

Table 3.1   Table of Laplace-transform pairs

	$F(s)$	$f(t) \quad 0 \le t$
1.	1	$u_1(t)$ unit impulse at $t = 0$
2.	$\dfrac{1}{s}$	1 or $u(t)$ unit step at $t = 0$
3.	$\dfrac{1}{s^2}$	$tu(t)$ ramp function
4.	$\dfrac{1}{s^n}$	$\dfrac{1}{(n-1)!}t^{n-1}$ n is a positive integer
5.	$\dfrac{1}{s}e^{-as}$	$u(t - a)$ unit step starting at $t = a$
6.	$\dfrac{1}{s}(1 - e^{-as})$	$u(t) - u(t - a)$ rectangular pulse
7.	$\dfrac{1}{s+a}$	e^{-at} exponential decay
8.	$\dfrac{1}{(s+a)^n}$	$\dfrac{1}{(n-1)!}t^{n-1}e^{-at}$ n is a positive integer
9.	$\dfrac{1}{s(s+a)}$	$\dfrac{1}{a}(1 - e^{-at})$
10.	$\dfrac{1}{s(s+a)(s+b)}$	$\dfrac{1}{ab}\left[1 - \dfrac{b}{b-a}e^{-at} + \dfrac{a}{b-a}e^{-bt}\right]$
11.	$\dfrac{s+\alpha}{s(s+a)(s+b)}$	$\dfrac{1}{ab}\left[\alpha - \dfrac{b(\alpha-a)}{b-a}e^{-at} + \dfrac{a(\alpha-b)}{b-a}e^{-bt}\right]$
12.	$\dfrac{1}{(s+a)(s+b)}$	$\dfrac{1}{b-a}(e^{-at} - e^{-bt})$
13.	$\dfrac{s}{(s+a)(s+b)}$	$\dfrac{1}{a-b}(ae^{-at} - be^{-bt})$
14.	$\dfrac{s+\alpha}{(s+a)(s+b)}$	$\dfrac{1}{b-a}[(\alpha - a)e^{-at} - (\alpha - b)e^{-bt}]$
15.	$\dfrac{1}{(s+a)(s+b)(s+c)}$	$\dfrac{e^{-at}}{(b-a)(c-a)} + \dfrac{e^{-bt}}{(c-b)(a-b)} + \dfrac{e^{-ct}}{(a-c)(b-c)}$
16.	$\dfrac{s+\alpha}{(s+a)(s+b)(s+c)}$	$\dfrac{(\alpha-a)e^{-at}}{(b-a)(c-a)} + \dfrac{(\alpha-b)e^{-bt}}{(c-b)(a-b)} + \dfrac{(\alpha-c)e^{-ct}}{(a-c)(b-c)}$
17.	$\dfrac{\omega}{s^2 + \omega^2}$	$\sin \omega t$
18.	$\dfrac{s}{s^2 + \omega^2}$	$\cos \omega t$
19.	$\dfrac{s+\alpha}{s^2 + \omega^2}$	$\dfrac{\sqrt{\alpha^2 + \omega^2}}{\omega}\sin(\omega t + \phi)$ $\phi = \tan^{-1}\dfrac{\omega}{\alpha}$
20.	$\dfrac{1}{s(s^2 + \omega^2)}$	$\dfrac{1}{\omega^2}(1 - \cos \omega t)$
21.	$\dfrac{s+\alpha}{s(s^2 + \omega^2)}$	$\dfrac{\alpha}{\omega^2} - \dfrac{\sqrt{\alpha^2 + \omega^2}}{\omega^2}\cos(\omega t + \phi)$ $\phi = \tan^{-1}\dfrac{\omega}{\alpha}$
22.	$\dfrac{1}{(s+a)(s^2 + \omega^2)}$	$\dfrac{e^{-at}}{a^2 + \omega^2} + \dfrac{1}{\omega\sqrt{a^2+\omega^2}}\sin(\omega t - \phi)$ $\phi = \tan^{-1}\dfrac{\omega}{\alpha}$
23.	$\dfrac{1}{(s+a)^2 + b^2}$	$\dfrac{1}{b}e^{-at}\sin bt$
24.	$\dfrac{1}{s^2 + 2\zeta\omega_n s + \omega_n^2}$	$\dfrac{1}{\omega_n\sqrt{1-\zeta^2}}e^{-\zeta\omega_n t}\sin \omega_n\sqrt{1-\zeta^2}\,t$
25.	$\dfrac{s+a}{(s+a)^2 + b^2}$	$e^{-at}\cos bt$
26.	$\dfrac{1}{s^2(s+a)}$	$\dfrac{1}{a^2}(at - 1 + e^{-at})$
27.	$\dfrac{1}{s(s+a)^2}$	$\dfrac{1}{a^2}(1 - e^{-at} - ate^{-at})$
28.	$\dfrac{s+\alpha}{s(s+a)^2}$	$\dfrac{1}{a^2}[\alpha - \alpha e^{-at} + a(a-\alpha)te^{-at}]$

CHAPTER 3

APPLICATIONS OF LAPLACE TRANSFORM

3.1 DEFINITION OF LAPLACE TRANSFORM

$$L[f(t)] = \int_0^\infty f(t)e^{-st}dt = F(s)$$

provided $f(t)$ is piecewise continuous over every finite interval and is of exponential order.

Laplace Transform of Some Simple Functions:

$L[\text{step function } u(t)] = \dfrac{1}{s}$ ... if $\sigma > 0$

$L[e^{-\alpha t}] = \dfrac{1}{s+\alpha}$ .... if $\sigma > -\alpha$

$L[\cos \omega t] = \dfrac{s}{s^2 + \omega^2}$ ...... if $\sigma > 0$

$L[\text{Ramp function } f(\dot{t}) = t] = \dfrac{1}{s^2}$ ... if $\sigma > 0$

where s is a complex variable and has the form $\sigma + j\omega$.

Table 3.1   Table of Laplace-transform pairs

	F(s)	f(t) $0 \leq t$
1.	1	$u_1(t)$ unit impulse at $t = 0$
2.	$\dfrac{1}{s}$	1 or $u(t)$ unit step at $t = 0$
3.	$\dfrac{1}{s^2}$	$tu(t)$ ramp function
4.	$\dfrac{1}{s^n}$	$\dfrac{1}{(n-1)!} t^{n-1}$ n is a positive integer
5.	$\dfrac{1}{s} e^{-as}$	$u(t-a)$ unit step starting at $t = a$
6.	$\dfrac{1}{s}(1 - e^{-as})$	$u(t) - u(t-a)$ rectangular pulse
7.	$\dfrac{1}{s+a}$	e^{-at} exponential decay
8.	$\dfrac{1}{(s+a)^n}$	$\dfrac{1}{(n-1)!} t^{n-1} e^{-at}$ n is a positive integer
9.	$\dfrac{1}{s(s+a)}$	$\dfrac{1}{a}(1 - e^{-at})$
10.	$\dfrac{1}{s(s+a)(s+b)}$	$\dfrac{1}{ab}\left[1 - \dfrac{b}{b-a} e^{-at} + \dfrac{a}{b-a} e^{-bt}\right]$
11.	$\dfrac{s+\alpha}{s(s+a)(s+b)}$	$\dfrac{1}{ab}\left[\alpha - \dfrac{b(\alpha-a)}{b-a} e^{-at} + \dfrac{a(\alpha-b)}{b-a} e^{-bt}\right]$
12.	$\dfrac{1}{(s+a)(s+b)}$	$\dfrac{1}{b-a}(e^{-at} - e^{-bt})$
13.	$\dfrac{s}{(s+a)(s+b)}$	$\dfrac{1}{a-b}(ae^{-at} - be^{-bt})$
14.	$\dfrac{s+\alpha}{(s+a)(s+b)}$	$\dfrac{1}{b-a}[(\alpha-a)e^{-at} - (\alpha-b)e^{-bt}]$
15.	$\dfrac{1}{(s+a)(s+b)(s+c)}$	$\dfrac{e^{-at}}{(b-a)(c-a)} + \dfrac{e^{-bt}}{(c-b)(a-b)} + \dfrac{e^{-ct}}{(a-c)(b-c)}$
16.	$\dfrac{s+\alpha}{(s+a)(s+b)(s+c)}$	$\dfrac{(\alpha-a)e^{-at}}{(b-a)(c-a)} + \dfrac{(\alpha-b)e^{-bt}}{(c-b)(a-b)} + \dfrac{(\alpha-c)e^{-ct}}{(a-c)(b-c)}$
17.	$\dfrac{\omega}{s^2+\omega^2}$	$\sin \omega t$
18.	$\dfrac{s}{s^2+\omega^2}$	$\cos \omega t$
19.	$\dfrac{s+\alpha}{s^2+\omega^2}$	$\dfrac{\sqrt{\alpha^2+\omega^2}}{\omega} \sin(\omega t + \phi)$ $\phi = \tan^{-1} \dfrac{\omega}{\alpha}$
20.	$\dfrac{1}{s(s^2+\omega^2)}$	$\dfrac{1}{\omega^2}(1 - \cos \omega t)$
21.	$\dfrac{s+\alpha}{s(s^2+\omega^2)}$	$\dfrac{\alpha}{\omega^2} - \dfrac{\sqrt{\alpha^2+\omega^2}}{\omega^2} \cos(\omega t + \phi)$ $\phi = \tan^{-1} \dfrac{\omega}{\alpha}$
22.	$\dfrac{1}{(s+a)(s^2+\omega^2)}$	$\dfrac{e^{-at}}{a^2+\omega^2} + \dfrac{1}{\omega\sqrt{a^2+\omega^2}} \sin(\omega t - \phi)$ $\phi = \tan^{-1} \dfrac{\omega}{\alpha}$
23.	$\dfrac{1}{(s+a)^2+b^2}$	$\dfrac{1}{b} e^{-at} \sin bt$
24.	$\dfrac{1}{s^2+2\zeta\omega_n s+\omega_n^2}$	$\dfrac{1}{\omega_n\sqrt{1-\zeta^2}} e^{-\zeta\omega_n t} \sin \omega_n\sqrt{1-\zeta^2}\, t$
25.	$\dfrac{s+a}{(s+a)^2+b^2}$	$e^{-at} \cos bt$
26.	$\dfrac{1}{s^2(s+a)}$	$\dfrac{1}{a^2}(at - 1 + e^{-at})$
27.	$\dfrac{1}{s(s+a)^2}$	$\dfrac{1}{a^2}(1 - e^{-at} - ate^{-at})$
28.	$\dfrac{s+\alpha}{s(s+a)^2}$	$\dfrac{1}{a^2}[\alpha - \alpha e^{-at} + a(a-\alpha)te^{-at}]$

Step 1) $AD^2 x_2 + MDx_2 + kx_2 = kx_1 \ldots$

the differential equation where $x_1(t)$ is the input and $x_2(t)$ is called the response function. First of all take the Laplace transform of both sides:

$$L[AD^2 x_2 + MDx_2 + kx_2] = L[kx_1].$$

2) Then take the Laplace transform of each term; after substituting in the original equation, rearrange the equation:

$$kx_1(s) = (As^2 + Ms + k)x_2(s)$$
$$- [Asx_2(o) + ADx_2(o) + Mx_2(o)]$$

characteristic function $x_2(s)$ - the transform equation

$$x_2(s) = \frac{kx_1(s) + Asx_2(o) + Mx_2(o) + ADx_2(o)}{As^2 + Ms + k}$$

3) $x_2(t) = L^{-1}[x_2(s)] \ldots$ the response function

3.3 INVERSE TRANSFORM

$$F(s) = P(s)/Q(s) = [a_n s^n + a_{n-1} s^{n-1} + \ldots]/[s^m + b_{m-1} s^{m-1}$$
$$+ \ldots]$$

$$= P(s)/[(s - s_1)(s - s_2) \ldots (s - s_m)]$$

after breaking $Q(s)$ into linear and quadratic factors.

Application of partial expansion in finding inverse Laplace transform.

A) $F(s)$ has first order real poles:

$$F(s) = \frac{P(s)}{Q(s)} = \frac{P(s)}{s(s-s_1)(s-s_2)} = \frac{A_0}{s} + \frac{A_1}{s-s_1} + \frac{A_2}{s-s_2}$$

$$\boxed{f(t) = L^{-1}[F(s)] = A_0 + A_1 e^{s_1 t} + A_2 e^{s_2 t}}$$

Two inverse transforms.

Evaluation of coefficients A_k:

$$A_k = \left[(s - s_k)\frac{P(s)}{Q(s)}\right]_{s=s_k}$$

A_k is called the residue of $F(s)$.

B) Multiple-order real poles:

$$F(s) = \frac{P(s)}{Q(s)} = \frac{P(s)}{(s-s_1)^2(s-s_2)} = \frac{A_{12}}{(s-s_1)^2} + \frac{A_{11}}{s-s_1} + \frac{A_2}{s-s_2}$$

$$f(t) = A_{12} \cdot t \cdot e^{s_1 t} + A_{11} \cdot e^{s_1 t} + A_2 e^{s_2 t}$$

A general formula for finding coefficients associated with the repeated real pole of order n,

$$A_{rn}\left[(s - s_r)^n \frac{P(s)}{Q(s)}\right]_{s=s_r}$$

$$A_{r(n-k)} = \frac{1}{k!}\frac{d^k}{ds^k}\left[(s - s_r)^n \frac{P(s)}{Q(s)}\right]_{s=s_r}$$

C) Complex conjugate poles:

$$f(s) = \frac{P(s)}{Q(s)} = \frac{P(s)}{(s^2 + 2\zeta\omega_n s + \omega_n^2)(s-s_3)}$$

$$= \frac{A_1}{s - s_1} + \frac{A_2}{s - s_2} + \frac{A_3}{s - s_3}$$

$$= \frac{A_1}{s + \zeta\omega_n - j\omega_n\sqrt{1 - \zeta^2}} + \frac{A_2}{s + \zeta\omega_n + j\omega_n\sqrt{1 - \zeta^2}}$$

$$+ \frac{A_3}{s - s_3}$$

Hence, the inverse function of $F(s)$ is,

$$f(t) = A_1 e^{(-\zeta\omega_n + j\omega_n\sqrt{1 - \zeta^2})t} + A_2 e^{(-\zeta\omega_n - j\omega_n\sqrt{1 - \zeta^2})t} + A_3 e^{s_3 t}$$

$$= 2|A_1| e^{6t} \sin(\omega_d t + \phi) + A_3 e^{s_3 t}$$

where ϕ = angle of $A_1 + 90°$

Now, $A_1 = [(s - s_1)F(s)]_{s=s_1}$ and $A_3 = [(s - s_3)F(s)]_{s=s_3}$

HAZONY AND RILEY RULE

For a normalized ratio of polynomials:

A) If the denominator is one degree higher than the numerator, the sum of the residues is one.

B) If the denominator is two or more degrees higher than the numerator, the sum of the residues is zero.

GRAPHICAL METHOD

$$F(s) = \frac{P(s)}{Q(s)} = k \frac{\prod_{m=1}^{\omega}(s - z_m)}{\prod_{k=1}^{\nu}(s - p_k)} = \frac{A_1}{s - p_1} + \frac{A_2}{s - p_2}$$

Pole-zero plot of the function $F(s)$:

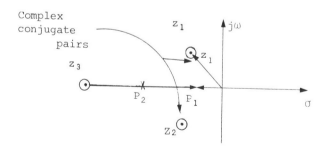

$A_k = k \frac{\text{Product of directed distances from each zero to the pole } p_k}{\text{Product of directed distances from all other poles to the pole } p_k}$

... except f $\sim$ k = 0.

An example:

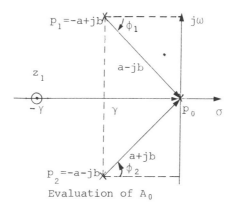

Evaluation of A_0

$$-F(s) = \frac{k(s+\gamma)}{s[(s+a)^2+b^2]} = \frac{k(s+\gamma)}{s(s+a-jb)(s+a+jb)} = \frac{k(s-z_1)}{s(s-p_1)(s-p_2)}$$

$$= \frac{A_0}{s} + \frac{A_1}{s+a-jb} + \frac{A_2}{s+a+jb}$$

$$A_0 = \frac{k\gamma}{(a-jb)(a+jb)} = \frac{k\gamma}{a^2+b^2}$$

$$A_1 = \frac{k[(\gamma-a)+jb]}{(-a+jb)(j2b)}$$

$$= \frac{k}{2b}\sqrt{\frac{(\gamma-a)^2+b^2}{a^2+b^2}} \; e^{j(\phi-\theta-\pi/2)}$$

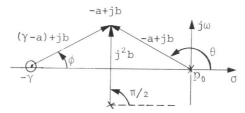

Fig. Evaluation of A_1.

where $\phi = \tan^{-1}\frac{b}{\gamma-a}$

$\theta = \tan^{-1}\frac{b}{-a}$

A_1 and A_2 are complex conjugate

$$\sigma - f(t) = A_0 + 2|A_1| e^{-\gamma t} \cdot \cos(bt + \phi - \theta - \pi/2)$$

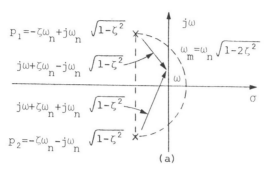

(a)

Pole-zero diagram.

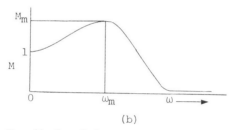

(b)

Magnitude of frequency response.

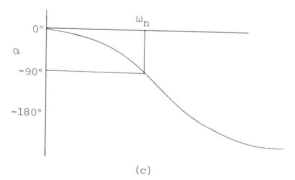

(c)

Angle of frequency response.

3.4 FREQUENCY RESPONSE FROM THE POLE-ZERO DIAGRAM

Frequency response: It is the steady-state response with a sine-wave forcing function for all values of frequency.

It is given by two curves:

A) M curve: the ratio of output amplitude to input amplitude as a function of frequency.

B) α curve: phase angle of the output α as a function of frequency.

Frequency-response characteristics:

A) at $\omega = 0$, the magnitude is a finite value and the angle is 0°.

B) If the number of poles is more than the number of zeros, as $\omega \to \infty$, the magnitude approaches zero and the angle is -90° times the difference between the number of poles and zeros.

C) In order to have a peak value M_m, there must be present complex poles near the imaginary axis. More precisely, the damping ratio must be less than 0.707.

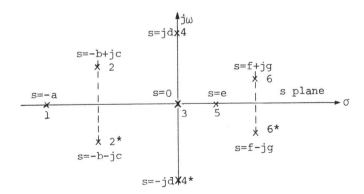

Location of poles in the s plane.

Table 3.1

Pole	Response	Characteristics
1	Ae^{-at}	Damped exponential
2-2*	$Ae^{-bt}\sin(ct + \phi)$	Exponentially damped sinusoid
3	A	Constant
4-4*	$A\sin(dt + \phi)$	Constant sinusoid
5	Ae^{et}	Increasing exponential (unstable)
6-6*	$Ae^{ft}\sin(gt + \phi)$	Exponentially increasing sinusoid (unstable)

3.5 ROUTH'S STABILITY CRITERION

Consider the characteristic equation

$$Q(s) = b_n s^n + b_{n-1} s^{n-1} + b_{n-2} s^{n-2} + \ldots + b_1 s + b_0 = 0$$

where all coefficients are real.

A) All powers of s from s^n to s^o must be present in the characteristic equation. The system is unstable if any coefficients other than b_o are zero or if any of the coefficients are negative. In this case, the roots are either imaginary or complex with positive real parts.

B) To determine the number of roots in the right half s plane:

The coefficients of the characteristic are arranged in the following Routhian array.

s^n	b_n	b_{n-2}	b_{n-4}	b_{n-6} ...
s^{n-1}	b_{n-1}	b_{n-3}	b_{n-5}	b_{n-7} ...
s^{n-2}	c_1	c_2	c_3	...
s^{n-3}	d_1	d_2	...	
.				
.				
.				
s^1	j_1			
s^0	k_1			

c_1, c_2, c_3, etc., are evaluated as follows:

$$c_1 = \frac{b_{n-1} b_{n-2} - b_n b_{n-3}}{b_{n-1}}$$

$$c_2 = \frac{b_{n-1} b_{n-4} - b_n b_{n-5}}{b_{n-1}}$$

$$c_3 = \frac{b_{n-1} b_{n-6} - b_n b_{n-7}}{b_{n-1}}$$

This pattern is continued until the rest of the c's are all equal to zero. The d row is formed by using the s^{m-1} and s^{m-2} row. The constants are

$$d_1 = \frac{c_1 b_{n-3} - b_{n-1} c_2}{c_1}$$

$$d_2 = \frac{c_1 b_{n-5} - b_{n-1} c_3}{c_1}$$

$$d_3 = \frac{c_1 b_{n-7} - b_{n-1} c_4}{c_1}$$

When no more d terms are present, the other rows are formed in a similar way.

Routh's criterion: The number of roots of the characteristic equation with positive real parts is equal to the number of changes of sign of the coefficients in the first column of the Routhian array. Thus a system will be stable if all terms in the first column have identical signs.

An example

The Routh's array

s^4	1	2	9
s^3	10	3	
s^2	$\frac{17}{10}$	9	
s^1	-50		
s^0	9		

There are two changes of sign, so there are two roots in the right-half of the s plane.

C) Theorems: (For special cases)

Division of a row: The coefficients of any row may be multiplied or divided by the number without changing the signs of the first column.

A zero coefficient in the first column: The procedure below can be used whenever the first term in a row is zero, but the other terms are not equal to zero.

Method A: A small positive number δ is substituted for the zero. The rest of the terms in the array are evaluated as usual.

Method B: Substitute in the original equation $s = \frac{1}{x}$; then solve for the roots of x with positive real parts. The number of roots x with positive real parts will be the same as the number of s roots with positive real parts.

A ZERO ROW

A) An auxiliary equation is formed from the preceding row.

B) The array is completed by replacing the all-zero row by the coefficients obtained by differentiating the auxiliary equation.

C) The roots of the auxiliary equation are also roots of the original equation. These roots occur in pairs and are the negatives of each other.

3.6 IMPULSE FUNCTION: LAPLACE-TRANSFORM AND ITS RESPONSE

$$f(t) = \frac{u(t) - u(t-a)}{a}$$ Analytical expression

$$f(s) = \frac{1 - e^{-as}}{as}$$ its Laplace-transform

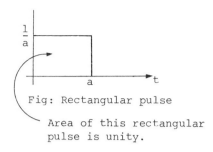

Fig: Rectangular pulse

Area of this rectangular pulse is unity.

A unit impulse: The limit of f(t) as $a \to 0$ is termed a unit impulse and is designated $b\delta(t)$.

$$\delta(t) = \lim_{a \to 0} f(t) = \lim_{a \to 0} \frac{u(t) - u(t-a)}{a}$$

$\Delta(s)$ = The Laplace-transform of the unit impulse = 1.

SECOND-ORDER SYSTEM WITH IMPULSE EXCITATION

$$F(s) = \frac{\omega_n^2}{s^2 + 2\zeta\omega_n s + \omega_2^2}$$ Laplace-transform with an impulse input

---

The impulse function is the derivative of the step function, so the response to an impulse is the derivative of the response to a step function.

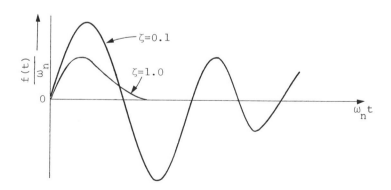

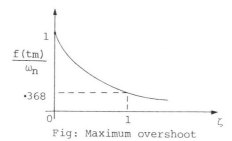

Fig: Maximum overshoot

$$t_m = \frac{\cos^{-1}\zeta}{\omega_n \sqrt{1-\zeta^2}}$$

where the maximum overshoot occurs.

$$f(t_m) = \omega_n \, e^{-(\zeta \cos^{-1}\zeta)/\sqrt{1-\zeta^2}}$$

CHAPTER 4

MATRIX ALGEBRA AND Z-TRANSFORM

4.1 FUNDAMENTALS OF MATRIX ALGEBRA

OPERATIONS WITH MATRICES

A) Matrix equality:

$\bar{A} = \bar{B}$ if and only if $a_{jk} = b_{jk}$ for $1 \leq j \leq p$ and $1 \leq k \leq q$ where the size of both matrices is $p \times q$.

B) Addition and subtraction:

If $\bar{A} = [a_{jk}]$ and $\bar{B} = [b_{jk}]$ are both $p \times q$ matrices, then $\bar{A} + \bar{B} = \bar{C}$ and $\bar{A} - \bar{B} = \bar{D}$ implies that $\bar{C} = [c_{jk}]$ and $\bar{D} = [d_{jk}]$ are also of the size $p \times q$ and $c_{jk} = a_{jk} + b_{jk}$ and $d_{jk} = a_{jk} - b_{jk}$ where $j = 1, 2, \ldots, p$ and $k = 1, 2, 3, \ldots, q$.

C) Matrix multiplication:

Scalar multiplication:

$$a\bar{B} = \bar{B}a = [ab_{jk}]$$

where a is a scalar.

Matrix multiplication:

$\bar{A} = [a_{jk}]$ of size $p \times q$ and $\bar{B} = [b_{jk}]$ of size $l \times m$. The product $\bar{A}\bar{B} = \bar{D}$ is defined only when $q = l$, then $\bar{A}$ and $\bar{B}$ are conformable.

$$d_{jk} = \sum_{k'=1}^{a} a_{jk'} b_{k'k}, \text{ size of } \bar{D} \text{ is } p \times m.$$

D) Transpose, conjugate and the associate matrix:

If $\bar{B} = [b_{jk}]$ then the transpose of $\bar{B}$ is $\bar{B}^T = [b_{kj}]$. The matrix $\bar{B}$ is symmetric if $\bar{B} = \bar{B}^T$. If $\bar{B} = -\bar{B}^T$, then $\bar{B}$ is skew-symmetric. $(\bar{A}\bar{B})^T = B^T A^T$.

$A^* = $ conjugate of $A = [\bar{a}_{jk}]$

Associate matrix of $\bar{A}$ = conjugate transpose of $\bar{A}$.

Hermitian matrices: These are the matrices for which $\bar{B} = \overline{B^T}$, these are called skew-hermitian if $\bar{A} = -\overline{A^T}$.

E) Matrix inversion:

$\bar{B}\bar{A} = \bar{A}\bar{B} = \bar{I}$ then $\bar{A} = \bar{B}^{-1}$...A should be a square matrix.

$A^{-1} = \dfrac{\bar{C}^T}{|\bar{A}|}$ where $\bar{C}^T$ is the adjoint matrix denoted by $\text{Adj}(\bar{A})$.

$A^{-1} = \text{Adj}(\bar{A})/|\bar{A}|$ Inverse of a non-singular matrix.

F) Special relationships:

$(\bar{A}\bar{B}\bar{C}\ldots\bar{F})^{-1} = \bar{F}^{-1}\ldots\bar{C}^{-1}\bar{B}^{-1}\bar{A}^{-1}$

If $\bar{A}^{-1} = A$, then $\bar{A}$ is called involutory.

If $\bar{B}^{-1} = \bar{B}^T$ then $\bar{B}$ is called orthogonal.

If $\bar{C}^{-1} = \bar{C}^{T*}$, then $\bar{C}$ is called unitary.

G) Cofactors and minors:

Minors: $\bar{B}$ is a $m \times m$ matrix, then minor M_{jk} is the determinant of $m_{-1} \times m_{-1}$ matrix formed from $\bar{B}$ by eliminating the $j + n$ row and the kth column.

Cofactors:

The cofactors are given by $C_{jk} = (-1)^{j+k} M_{jk}$ where every element b_{jk} of $\bar{B}$ has a cofactor C_{jk}.

Determinants:

$\bar{B}$ is a m × n matrix, then $|\bar{B}| = \sum_{j=1}^{m} b_{ij} C_{ij}$ where i is any arbitrary row.

The Laplace expansion can be done with respect to any column j' to get $|\bar{B}| = \sum_{i=1}^{} b_{ij'} c_{ij'}$

H) Integration and differentiation of matrices:

If $\bar{C}(t) = [c_{jk}(t)]$, then $\dfrac{d\bar{c}}{dt} = \left[\dfrac{d}{dt} c_{jk}(t)\right]$ and

$$\int \bar{D}\, dt = \left[\int d_{ij}(t) dt\right]$$

Useful role for differentiating a determinant:

$$\dfrac{\partial |\bar{B}|}{\partial b_{ij}} = \bar{C}_{ij}$$

PARTITIONED MATRICES

$\bar{C}\, \bar{D} = \bar{E}$ can be divided as follows:

$$\left[\begin{array}{c}\bar{C}_1 \\ \hline \bar{C}_2\end{array}\right] [D_1 \mid D_2] = \left[\begin{array}{c|c}\bar{C}_1 \bar{D}_1 & \bar{C}_1 \bar{D}_2 \\ \hline \bar{C}_2 \bar{D}_1 & \bar{C}_2 \bar{D}_2\end{array}\right] = \left[\begin{array}{c|c}\bar{F}_1 & \bar{F}_2 \\ \hline \bar{F}_3 & \bar{F}_4\end{array}\right]$$

$$\left[\begin{array}{c|c}\bar{C}_1 & \bar{C}_2 \\ \hline \bar{C}_3 & \bar{C}_4\end{array}\right]\left[\begin{array}{c}D_1 \\ \hline D_2\end{array}\right] = \left[\begin{array}{c}\bar{C}_1 \bar{D}_1 + \bar{C}_2 \bar{D}_2 \\ \hline \bar{C}_3 \bar{D}_1 + \bar{C}_4 \bar{D}_2\end{array}\right] = \left[\begin{array}{c}E_1 \\ \hline E_2\end{array}\right]$$

A position can be used to find the inverse of a matrix $\bar{C}$.

$$\bar{C}\, \bar{D} = \bar{I} \rightarrow \left[\begin{array}{c|c}\bar{C}_1 & \bar{C}_2 \\ \hline \bar{C}_3 & \bar{C}_4\end{array}\right]\left[\begin{array}{c|c}\bar{D}_1 & \bar{D}_2 \\ \hline \bar{D}_3 & \bar{D}_4\end{array}\right] = \left[\begin{array}{c|c}I & O \\ \hline O & I\end{array}\right]$$

This position results in two simultaneous equations $\bar{C}_1 \bar{D}_1 + \bar{C}_2 \bar{D}_3 = I$ and $\bar{C}_3 \bar{D}_1 + \bar{C}_4 \bar{D}_3 = 0$ from which $\bar{D}_1$ and $\bar{D}_3$ can be obtained. Two other equations give $\bar{D}_2$ and $\bar{D}_4$ and ultimately result into $\bar{C}^{-1}$.

Diagonal, block diagonal and triangular matrices:

Diagonal matrix: If all the non-zero elements of a square matrix are on the main diagonal, then it is called a diagonal matrix.

Triangular matrix: A square matrix, all of whose elements below (or above) the main diagonal are equal to zero, is called an upper (or lower) triangular matrix.

4.2 Z-TRANSFORMS

This is especially useful for linear systems with sampled or discrete input signals.

Block diagram:

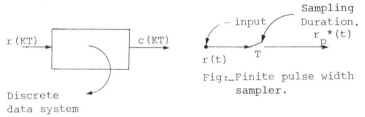

Fig:_Finite pulse width sampler.

The input for a discrete data system is given by

$$r^*(t) = \sum_{k=0}^{\infty} r(kT) \delta(t - kT)$$

where k varies from 0 to ∞ and these numbers are T seconds apart.

The output of the sampler is given by:

$$r_p^*(t) = r(t) \sum_{k=0}^{\infty} [u(t - kT) - u(t - kT - p)]$$

where u is the unit step function, T is the sampling period, and P is the sampling ration ($P \ll T$).

DEFINITION

The z-transform is defined as $z = e^{Ts}$, where s is the Laplace transform variable.

47

$$s = \frac{1}{T} \ln z$$

Table of z-transforms:

Table 4.1

Laplace Transform	Time Function	z-Transform
1	Unit impulse $\delta(t)$	1
$\frac{1}{s}$	Unit step $u(t)$	$\frac{z}{z-1}$
$\frac{1}{1-e^{-Ts}}$	$\delta_T(t) = \sum_{n=0}^{\infty} \delta(t-nT)$	$\frac{z}{z-1}$
$\frac{1}{s^2}$	t	$\frac{Tz}{(z-1)^2}$
$\frac{1}{s^3}$	$\frac{t^2}{2}$	$\frac{T^2 z(z+1)}{2(z-1)^3}$
$\frac{1}{s^{n+1}}$	$\frac{t^n}{n!}$	$\lim_{a \to 0} \frac{(-1)^n}{n!} \frac{\partial^n}{\partial a^n} \left(\frac{z}{z - e^{-aT}} \right)$
$\frac{1}{s+a}$	e^{-at}	$\frac{z}{z - e^{-aT}}$
$\frac{1}{(s+a)^2}$	te^{-at}	$\frac{Tze^{-aT}}{(z - e^{-aT})^2}$
$\frac{a}{s(s+a)}$	$1 - e^{-at}$	$\frac{(1 - e^{-aT})z}{(z-1)(z - e^{-aT})}$
$\frac{\omega}{s^2 + \omega^2}$	$\sin \omega t$	$\frac{z \sin \omega T}{z^2 - 2z \cos \omega T + 1}$
$\frac{\omega}{(s+a)^2 + \omega^2}$	$e^{-at} \sin \omega t$	$\frac{ze^{-aT} \sin \omega T}{z^2 e^{2aT} - 2ze^{aT} \cos \omega T + 1}$
$\frac{s}{s^2 + \omega^2}$	$\cos \omega t$	$\frac{z(z - \cos \omega T)}{z^2 - 2z \cos \omega T + 1}$
$\frac{s+a}{(s+a)^2 + \omega^2}$	$e^{-at} \cos \omega t$	$\frac{z^2 - ze^{-aT} \cos \omega T}{z^2 - 2ze^{-aT} \cos \omega T + e^{-2aT}}$

INVERSE Z

Methods:

A) Partial-fraction expansion

B) Power-series

C) The inversion formula

POWER-SERIES METHOD

A) The z-transform is expanded into a power series in powers of z^{-1}.

B) The coefficient of z^{-k} is the value of $r(t)$ at $t = kT$.

INVERSION FORMULA

Inversion formula: $r(kT) = \dfrac{1}{2\pi j} \oint_L R(z) z^{k-1} dz$

where L is a circle of radius $|z| = t^{cT}$. The center of this is at the origin and c is such that all the poles of $R(z)$ are inside the circle.

THEOREMS

A) Addition and Subtraction

If $r_1(kT)$ and $r_2(kT)$ have z-transforms $R_1(z)$ and $R_2(z)$, respectively, then

$$\zeta[r_1(kT) \pm r_2(kT)] = R_1(z) \pm R_2(z)$$

B) Multiplication by a Constant

$$\zeta[ar(kT)] = a\zeta[r(kT)] = aR(z)$$

where a is a constant.

C) Real Translation

$$\zeta[r(kT - nT)] = z^{-n} R(z)$$

and

$$\zeta[r(kT + nT)] = z^n \left[R(z) - \sum_{k=0}^{n-1} r(kT) z^{-k} \right]$$

D) Complex Translation

$$\zeta[e^{\mp akT} r(kT)] = R(ze^{\pm aT})$$

E) Initial-Value Theorem

$$\lim_{k \to 0} r(kT) = \lim_{z \to \infty} R(z)$$

F) Final-value Theorem

$$\lim_{k \to \infty} r(kT) = \lim_{z \to 1} (1 - z^{-1})R(z)$$

if the function, $(1 - z^{-1})R(Z)$, has no poles on or outside the circle centered at the origin in the z-plane, $|z| = 1$.

CHAPTER 5

SYSTEM'S REPRESENTATION: BLOCK DIAGRAM, TRANSFER FUNCTIONS, AND SIGNAL FLOW GRAPHS

5.1 BLOCK DIAGRAM AND TRANSFER FUNCTION

Fundamentals:

A block diagram is a symbolic representation of:

A) the flow of information in a system, and

B) the functions performed by each component in the system.

The transfer function expresses the relationship between output and input of a system and is frequently given in terms of the Laplace variable s.

In finding the transfer function of a system, all initial conditions must be set to zero.

The transfer function may be expressed in terms of:

A) the operator D . . . . (D)

B) the Laplace-transform variable s . . . . C(j)

C) in phasor form . . . . C(jω)

This latter case is used in the sinusoidal steady-state analysis.

The denominator of the transfer function is the characteristic equation when it is set equal to zero.

Summation points in the block diagrams:

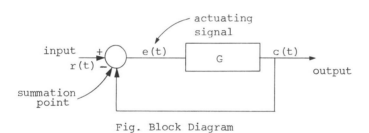

Fig. Block Diagram

Transfer function:

$$G(D) = \text{Transfer function} = \frac{V_{out}}{V_{in}}$$

$G(S)$ = Laplace-transform of the transfer function

$$= \frac{E_2(s)}{E_1(s)} = RCs/1 + RCs$$

$\overline{G}(j\omega)$ = Frequency transfer function = $\dfrac{\overline{E}_2(j\omega)}{\overline{E}_1(j\omega)}$

Blocks in Cascade:

If the operation of an element or a component can be described independently, then a block can be used to represent that component.

Combination of cascade blocks:

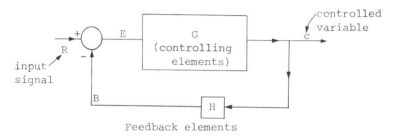

Fig. Equivalent Block diagram after cascading.

Determination of control ratio:

$$\boxed{\text{Control ratio} = \frac{C(s)}{R(s)} = \frac{G(s)}{1 + G(s)H(s)}\\ = \text{Close-loop transfer function}}$$

$$\boxed{\text{Characteristic equation:} \quad 1 + G(s)H(s) = 0}$$

$$\boxed{\begin{array}{l}\text{Open-loop} \\ \text{transfer function}\end{array} = \frac{B(s)}{E(s)} = G(s)H(s)}$$

$$\boxed{\begin{array}{l}\text{Forward transfer} \\ \text{function}\end{array} = \frac{C(s)}{E(s)} = G(s)}$$

5.2 TRANSFER FUNCTIONS OF THE COMPENSATING NETWORKS

Lag compensator:

$$\frac{E_1(s)}{E_2(s)} = \frac{1 + (R_1 + R_2)Cs}{1 + R_2 Cs}$$

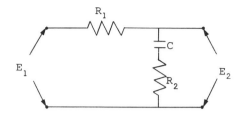

$$\frac{E_2(s)}{E_1(s)} = \frac{1 + R_2 Cs}{1 + (R_1 + R_2)Cs} = G(s)$$

$$G(s) = \frac{1}{\alpha} \frac{s + 1/T}{s + 1/\alpha T}$$

$\ldots \ \alpha = \dfrac{R_1 + R_2}{R_2}$ and $T = R_2 C$

Lead compensator:

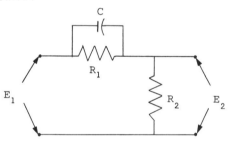

$$G(s) = \frac{s + 1/T}{s + 1/\alpha T} \qquad \alpha = \frac{R_2}{R_1 + R_2}$$

and $T = R_1 C$

Lag-lead compensator:

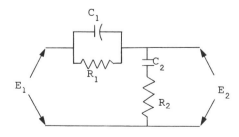

LAPLACE TRANSFORM THEOREMS

A) Linearity

$L[a \cdot f(t)] = a \cdot L[f(t)] = a \cdot F(s)$ ... if a is a constant

B) Superposition

$L[f_1(t) \pm f_2(t)] = F_1(s) \pm F_2(s)$

C) Translation in time

$L[f(t - a)] = e^{-as} F(s)$ ... if a is a positive real, and $f(t - a) = 0$ for $0 < t < a$.

D) Complex differentiation

$L[t \cdot f(t)] = \frac{-d}{ds} F(s)$

E) Translation in tne s domain

$L[e^{at} \cdot f(t)] = F(s - a)$

F) Real differentiation

$L[Df(t)] = sF(s) - f(o^+)$

$f(o^+)$ is the value of the limit of $f(t)$ as the origin, $t = 0$ is approached from the right-hand side.

$L[D^2 f(t)] = s^2 F(s) - sf(o) - Df(o)$

$L[D^n f(t)] = s^n F(s) - s^{n-1} f(o) \ldots - D^{n-1} f(o)$

G) Real integration

$L[D^{-n} f(t)] = \frac{F(s)}{s^n} + \frac{D^{-1} f(o)}{s^n} + \ldots + \frac{D^{-n} f(o)}{s}$

H) Final value F

if $f(t)$ and $DF(t)$ are Laplace transformable, then

$\lim_{s \to o} sf(s) = \lim_{t \to \infty} f(t)$

According to the final value theorem we can find the

final value of the function f(t) by working in the s domain which saves us the work involved in taking the inverse Laplace transform.

Limitations of this theorem:

A) Whenever SF(s) has poles on the imaginary axis or in the right half s plane, there is no finite final value of f(t) since SF(s) becomes infinite and the theorem cannot be used.

B) Suppose the driving function is sinusoidal and is equal to $\sin\omega t$. The $L[\sin\omega t]$ has poles at $s = \pm j\omega$; furthermore, $\lim_{t\to\infty} \sin\omega t$ does not exist. Thus this theorem is invalid whenever the driving function is sinusoidal.

INITIAL VALUE THEOREM

$$\lim_{s\to\infty} S F(S) = \lim_{t\to 0} f(t) \ldots \text{if } \lim S F(S) \text{ exists.}$$

COMPLEX INTEGRATION THEOREM

$$L\left[\frac{f(t)}{t}\right] = \int_{s}^{\infty} F(s)ds$$

providing the limit $f(t)/t$ exists
$\quad\quad\quad\quad\quad t\to 0^+$

In words this means that division by the variable in the real time domain is equivalent to integration with respect to s in the s domain.

3.2 APPLICATION OF LAPLACE TRANSFORM TO DIFFERENTIAL EQUATIONS

An example

$$G(s) = \frac{E_2(s)}{E_1(s)} = \frac{1 + (T_1 + T_2)s + T_1 T_2 s^2}{1 + (T_1 + T_2 + T_{12})s + T_1 T_2 s^2}$$

$$\ldots \quad T_1 = R_1 C_1, \quad T_2 = R_2 C_2 \quad T_{12} = R_1 C_2$$

$$= \frac{(1 + R_1 C_1 s)(1 + R_2 C_2 s)}{(1 + R_1 C_1 s)(1 + R_2 C_2 s) + R_1 C_2 s}$$

5.3 SIGNAL FLOW GRAPHS

This graph is the pictorial representation of a set of simultaneous equations. The nodes represent the system variable and a branch acts as a signal multiplier.

DEFINITION AND ALGEBRA

Node: A node performs two functions:

A) Addition of all the incoming signals.

B) Transmission or distribution of the total incoming signals of a node to all the outgoing branches.

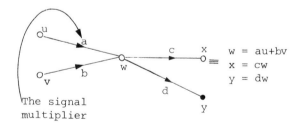

The signal multiplier

$w = au + bv$
$x = cw$
$y = dw$

These are the equations represented by the adjacent flow graph.

FLOW-GRAPH ALGEBRA
Rules

A) Series paths (cascade nodes): Series paths are combined into a single path by multiplying the transmittances of the individual paths.

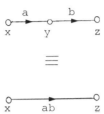

B) Parallel paths: Parallel paths are combined by adding the transmittance.

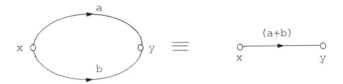

C) Elimination of a node: A node representing a variable can be eliminated as follows:

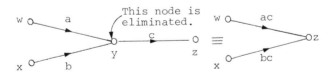

MASON'S RULE gives the overall transmittance of a system and is defined as follows:

$$T = \text{overall transmittance} = \frac{\Sigma T \Delta}{\Delta} \quad \text{where}$$

A) Δ represents the determinant of the graph given by

$$\Delta = 1 - \Sigma L_1 + \Sigma L_2 - \Sigma L_3 + \ldots \text{ and}$$

a) L_1 is defined as the transmittance of each loop, thus ΣL_1 would be the sum of the individual transmittances of all loops.

b) ΣL_2 = sum of L_2 where L_2 is the product of the transmittances of 2 non-touching loops. Two loops are non-touching if they do not share a common node.

c) ΣL_3 = sum of all possible combinations of the product of transmittances of non-touching loops taken three at a time.

B) T denotes the transmittance of each forward path between a source and a sink node.

C) If we remove the path which has transmittance T, the determinant of the subgraph produced is denoted by Δ.

CHAPTER 6

SERVO CHARACTERISTICS: TIME-DOMAIN ANALYSIS

6.1 TIME-DOMAIN ANALYSIS USING TYPICAL TEST SIGNALS

Following signals are used for analysis:

A) Step function: This function is defined as follows.

(i) Step Function

$$r(t) = \begin{cases} R_0 & \text{for } t > 0 \\ 0 & \text{for } t < 0 \end{cases}$$

$$= R_0 u(t)$$

where $u(t)$ is the unit step function.

B) Ramp function:

$$r(t) = R_0 t \quad \text{for } t \geq 0 \text{ and zero elsewhere}$$

$$= R_0 t u(t)$$

ii) Ramp Function

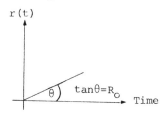

C) Parabolic function:

(iii) Parabolic Function

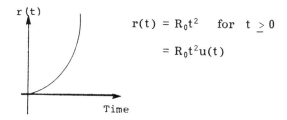

$$r(t) = R_0 t^2 \quad \text{for} \quad t \geq 0$$
$$= R_0 t^2 u(t)$$

Any random input signal can be considered as being composed of these signals.

CONCLUSION

If any linear system is analyzed mathematically or experimentally for each of these input signals then it can be said that the response of this system to these basic test signals will be the representation of the actual system response to any random signal.

6.2 TYPES OF FEEDBACK SYSTEMS

The standard form of the transfer function is

$$G(s) = \frac{k_n(1 + a_1 s + a_2 s^2 + \ldots + a_m s^m)}{s^n(1 + b_1 s + b_2 s^2 + \ldots + b_p s^p)}$$

where $a_1, a_2, \ldots, b_1, b_2, \ldots$ are constant coefficients.

k_n = overall gain of transfer function $G(s)$

$n = 0,1,2$ denotes the type of the transfer function.

Characteristics of different types of systems:

A) Type 0: A constant error signal $E(s)$ results in a constant value of the output signal $C(s)$.

B) Type 1: A constant $E(s)$ signal results in a constant rate of change of the output signal.

C) Type 2: A constant $E(s)$ signal will produce a constant D^2C of the output variable.

Note: A type of the given system can be readily known from $G(j\omega)$ and log $G(j\omega)$ versus ω plots.

6.3 GENERAL APPROACH TO EVALUTION OF ERROR

IMPORTANT FACTS

A) Final-value theorem:

$$\lim_{t \to \infty} f(t) = \lim_{s \to 0} sF(s)$$

B) Differentiation theorem:

$$L[D^n C(t)] = s^n C(s)$$

when all the initial conditions are zero.

C) If the input signal to a unit feedback, stable system is a power series, then the steady-state output will have the same form as the input.

Steady-state error function: General Form

$$G(s) = \frac{C(s)}{E(s)} = \frac{k_n(1 + T_1 s)(1 + T_2 s)\ldots}{s^n(1 + T_a s)(1 + T_b s)\ldots}$$

$$E(s) = \frac{(1 + T_a s)(1 + T_b s)\ldots s^n C(s)}{s^n(1 + T_a s)(1 + T_b s)(1 + T_c s)}$$

This is obvious after rearranging the top equation.

$$e(t)_{ss} = \lim_{s \to 0} [SE(s)] = \lim_{s \to 0} \left[\frac{s(1+T_a s)(1+T_b s)\ldots\ldots s^n C(s)}{k_n (1+T_1 s)(1+T_2 s)} \right]$$

$$= \lim_{s \to 0} \frac{s[s^n C(s)]}{K_n}$$

This is the steady-state error.

This can be written as follows:

$$\boxed{e(t)_{ss} = \frac{D^n C(t)_{ss}}{k_n}}$$

This equation is useful when $D^n C(t)_{ss}$ = constant. Then $e(t)_{ss}$ will be constant and is equal to E_o. So $k_n E_o = D^n C(t)_{ss}$ = constant = C_n.

$$E(s) = \frac{s^n(1 + T_a s)(1 + T_b s)\ldots R(s)}{s^n(1 + T_a s)(1 + T_b s) + \ldots + k_n(1 + T_1 s)(1 + T_2 s)}$$

When $H(s) = 1$, i.e. unity feedback.

$$\boxed{e(t)_{ss} = \lim_{s \to 0} s \left[\frac{s^n(1 + T_a s)(1 + T_b s)\ldots R(s)}{s^n(1+T_a s)(1+T_b s) + \ldots + k_n(1+T_1 s)(1+T_2 s)} \right]}$$

The general expression.

ERROR SERIES: CONCEPTS

This error series is useful when the input to a feedback system is an arbitrary function of time.

Mathematical representation:

$$E(s) = R(s)/1 + G(s)H(s)$$

$$e(t) = \int_{-\infty}^{t} A(\tau)r(t - \tau)d\tau$$

where $A(q)$ is the inverse Laplace transform of $1/(1 + G(s)H(s))$.

Error series:

A) $e(t) = r(t)\int_{0}^{t} A(\tau)d\tau - r(t)\int_{0}^{t} \tau A(\tau)d\tau$

$$+ r(t)\int_{0}^{t} \frac{\tau^2}{2!} A(\tau)d\tau$$

where $r(t - \tau) = r(t) - \tau r(t) + \frac{\tau^2}{2!} r(t)$ Taylor expansion

B) $e_s(t) = C_0 r_s(t) + C_1 r_s(t) + \frac{C_2}{2!} r_s(t)$

Where C_0, C_1 are called the error coefficients and $e_s(t)$ is known as the error series.

$$C_n = (-1)^n \int_{0}^{0} \tau^n A(\tau)d\tau$$

Evaluation of the error coefficients from $A(s)$:

A) $C_0 = \lim_{s \to 0} A(s) = \lim_{s \to 0} \int_{0}^{\infty} A(s)e^{-\tau s}d\tau$

B) $C_1 = \lim_{s \to 0} \frac{d}{ds} A(s)$

C) $C_n = \lim_{s \to 0} \frac{d^n}{ds^n} A(s)$

6.4 ANALYSIS OF SYSTEMS: UNITY FEEDBACK

TYPE 0 SYSTEM $(n = 0)$

Step input $(r(t) = R_0 u(t))$:

$$e(t)_{ss} = \frac{R_0}{1 + k_0} = \text{constant}$$
$$= E_0$$

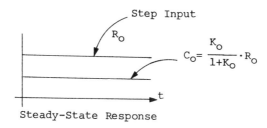

Steady-State Response

This is obtained from the general expression for $e(t)_{ss}$. Since

$$R(s) = \frac{R_0}{s}$$

Ramp input:

$$e(t)_{ss} = \infty$$

since $r(t) = R_1 t\, u(t)$

$$R(s) = R_1/s^2$$

Parabolic input:

$$e(t)_{ss} = r(t)_{ss} - C(t)_{ss} \quad \text{which approaches a value of infinity}$$
$$= \infty \quad \text{since } r(t) = R_2 t^2 u(t)$$

Conclusions:

A) A constant input (i.e. step input) produces a constant value of the output with a constant error signal.

B) When a ramp-function input produces a ramp output with a smaller slope, there is an error which approaches a value of ∞ with increasing time.

C) Type 0 system cannot follow a parabolic input.

TYPE 1 SYSTEM (n = 1)

Step input:

$$\boxed{e(t)_{ss} = 0}$$

This is obtained from the general equation by putting n = 1 and

$$R(s) = \frac{R_0}{s}$$

Ramp input:

$$\boxed{e(t)_{ss} = \infty}$$
since $r(t) = R_2 t^2 u(t)$
$R(s) = 2R_2/s^3$

Parabolic input:

$$\boxed{e(t)_{ss} = \frac{R_1}{k_1} = \text{constant} = E_0 \neq 0}$$
since $R(s) = R_1/s^2$

Conclusions:

A) Type 1 system with a constant input produces a

steady-state constant output of value equal to the input, i.e. zero steady-state error.

B) Type 1 system with a ramp input produces a ramp output with a constant error signal.

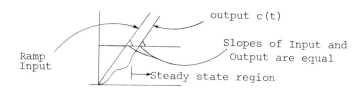

C) With a parabolic input, a parabolic output is produced with an error which increases with time $e(\infty) = \infty$.

TYPE 2 SYSTEM (n = 2)

Step input:

$$\boxed{e(t)_{ss} = 0}$$

This is obtained by putting $n = 2$ and $R(s) = R_0/s$.

Ramp input:

$$e(t)_{ss} = 0 \qquad \text{since } R(s) = R_1/s^2$$

Parabolic input:

$$e(t)_{ss} = \frac{2 \cdot R_2}{k_2} = \text{constant} \quad \text{since} \quad R(s) = \frac{2R_2}{s^3}$$

$$= E_0 \neq 0$$

Summary:

Table 6.1

Steady-state characteristics: Unity feedback systems

System type n	$r(t)_{ss}$ Steady-state (input)	(Steady-state error function)	(Steady-state output)	Value of $e(t)$ at $t = \infty$
0	Step	$\dfrac{R_0}{1+K_0}$	$\dfrac{K_0}{1+K_0} R_0$	$\dfrac{R_0}{1+K_0}$
0	Ramp	$\dfrac{R_1}{1+K_0} t - C_0$	$\dfrac{K_0 R_1}{1+K_0} t + C_0$	∞
0	Parabolic	$\dfrac{R_2}{1+K_0} t^2 - C_1 t - C_0$	$\dfrac{K_0 R_2}{1+K_0} t^2 + C_1 t + C_0$	∞
1	Step	0	R_0	0
1	Ramp	$\dfrac{R_1}{K_1}$	$R_1 t - \dfrac{R_1}{K_1}$	$\dfrac{R_1}{K_1}$
1	Parabolic	$-C_1 t - C_0$	$R_2 t_2 + C_1 t + C_0$	∞
2	Step	0	R_0	0
2	Ramp	0	$R_1 t$	0
2	Parabolic	$2\dfrac{R_2}{K_2}$	$R_2 t^2 - \dfrac{2R_2}{K_2}$	$\dfrac{2R_2}{K_2}$

CHAPTER 7

ROOT LOCUS

7.1 ROOTS OF CHARACTERISTIC EQUATIONS

Basic approach:

$$G(s) = \frac{C(s)}{E(s)} = \frac{k}{s(s+a)}$$

forward transfer function

STATIC LOOP SENSITIVITY

This is the value of k, when the transfer function is expressed in such form that the coefficients of s in both numerator and denominator are equal to unity.

$$\frac{C(s)}{R(s)} = \frac{k}{s^2 + 2s + k} = \frac{k}{s^2 + 2\zeta\omega_n s + \omega_n^2} \quad \text{for } a = 2$$

Roots of the characteristic equation:

$$s = -\zeta\omega_n \pm \omega_n\sqrt{\zeta^2 - 1}$$

where $\omega_n = \sqrt{k}$ and $\zeta = \dfrac{1}{\sqrt{k}}$

Plot of the roots of the characteristic equation:

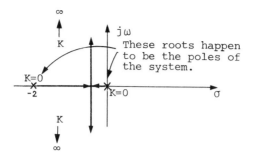

Roots are obtained for different values of k and these are plotted as shown.

A smooth curve is drawn (as shown by thick lines) through those points.

System performace v/s sensitivity k: salient points from the root locus.

An increase in the gain of the system results in:

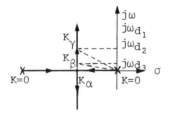

Fig. A sample plot.

A) a decrease in the damping ratio ζ, so the overshoot of the time response increases.

B) increase in ω_n (i.e., undamped natural frequency).

C) increase in ω_d (i.e., damped natural frequency)... ω_d is the imaginary component of the complex root.

D) The rate of decay, σ, is unchanged.

E) Root locus is a vertical line for $k \geq k_\alpha$ and

$$\zeta \omega_n = \sigma = \text{const.}$$

7.2 IMPORTANT PROPERTIES OF THE ROOT LOCI

EFFECT OF ADDITION OF POLES

When a pole is added to the function $G(s)H(s)$ in the left half of the s-plane, the net effect is of pushing the original locus towards the right half plane.

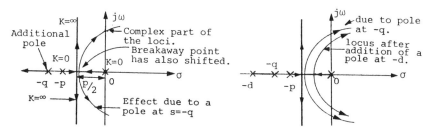

Fig. Original plot   Fig. Addition of pole at -d

ADDITION OF ZERO

Adding zeros has the effect of moving the root locus towards the left half of the s-plane.

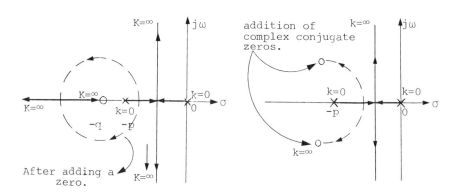

STEPS IN PLOTTING THE ROOT-LOCUS

A) The open loop transfer function $G(s)H(s)$ of the system is first determined.

B) The numerator and the denominator of the transfer function are then factorized into factors of the form (s + p).

C) Locate zeros and poles of the open-loop transfer function in the s-plane.

D) Determine the roots of the characteristic of the closed loop system, i.e. $[1 = G(s)H(s) = 0]$. The locus of the roots of the characteristic equation is then found from the plotted zeros and poles of the open loop function.

E) The locus is calibrated in terms of k. If the gain of the open-loop system is predetermined, the location of the exact roots of HG(s)H(s) is known. If the location of the roots is specified, k can be known.

F) Now that the roots are known, the system's response is the inverse Laplace transform.

G) If the specifications are not satisfied, then the shape that meets the desired specifications is determined and compensating networks are introduced in the system to meet these requirements.

Poles of the control ratio $\frac{C(s)}{R(s)}$:

$$\boxed{\frac{C(s)}{R(s)} = M(s) = \frac{N_1 D_2}{D_1 D_2 + N_1 N_2} = \frac{P(s)}{Q(s)}}$$

where $G(s) = \frac{N_1(s)}{D_1(s)}$, $H(s) = \frac{N_2(s)}{D_2(s)}$

and $B(s) = 1 + G(s)H(s) = D_1 D_2 + N_1 N_2 / D_1 D_2$

i.e. the zeros of B(s) are the poles of $\frac{C(s)}{R(s)}$ and these zeros determine the transient response.

Factors of Q(s) fall into the following categories:

Pole of C(s)	Corresponding inverse of the form u(t) - unit step
S	
$S + \frac{1}{T}$	$e^{-t/T}$
$s^2 + 2\zeta\omega_n s + \omega_n^2$	$e^{-\zeta\omega_n t} \sin(\omega_n \sqrt{1-\zeta^2}\, t + \phi),\ \zeta < 1$

P(s) only modifies the constant multiplier of these transients.

Conditions to plot the root-locus for k > 0:

$|G(s)H(s)| = 1$ Magnitude condition

$\underline{/G(s)H(s)} = (1 + 2m)180°$ for $m = 0, \pm 1, \pm 2 \ldots$ for $k > 0$.

Angle condition

Conditions for negative values of k:

$|G(s)H(s)| = 1$ Magnitude condition

$\underline{/G(s)H(s)} = (m)360° \ldots$ for $m = 0, \pm 1, \pm 2 \ldots$ for $k < 0$

Angle condition

The magnitude and angle conditions:

APPLICATION

$$G(s)H(s) = \frac{k(s - z_1)}{s^n(s - p_1)(s - p_2)}$$ General form of the open-loop transfer function.

Note: $s - p_1, s - p_2, \ldots, s - z_1$ etc., are the complex numbers representing the directed line segments.

$1 + G(s)H(s) = 0$ The characteristic equation.

Application of the magnitude and angle conditions results in

$$\frac{|k||s - z_1| \ldots |s - z_n|}{|s^n||s - p_1||s - p_2| \ldots} = 1$$

$$-\zeta = \underline{/s - z_1} \ldots + \underline{/s - z_w} - n\underline{/s} - \underline{/s - p_1}$$

$$= \begin{cases} (1 + 2m)180° & \ldots \text{ if } k > 0 \\ (m)360° & \ldots \ldots \text{ if } k < 0 \end{cases}$$

So modification of these equations results in:

A) $|k|$ = static loop sensitivity = $\dfrac{|s^n||s - p_1| \ldots |s - p_q|}{|s - z_1| \ldots |s - z_m|}$

B) $+\zeta$ = -(angles of numerator terms) + (angles of denominator terms)

$$= \begin{cases} -(1+2m)180° & \text{...if } k > 0 \\ -(m)360° & \text{........if } k < 0 \end{cases}$$

Note: These two conditions are used in the graphical construction of the root locus.

Construction rules for plotting the root locus:

A) Total number of branches in the complete root loci:

The number of branches in a root loci is equal to the number of poles of the open-loop transfer function.

If the equation is expressed in the following form

$$s^q + b_1 \ldots s^{q-1} \ldots b_n + k(s^p + a_1 s^{p-1} \ldots) = 0$$

then the number of branches of the root loci is greater or equal to q and p.

B) Locus on the real axis:

If the total number of poles and zeros to the right of the search point on the real axis is odd, then this point is on the root locus.

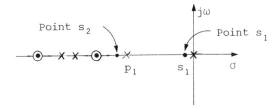

Note: Point s_1 lies on the real axis root locus but not s_2.

The following is the required angle condition so that the selected search point will be on the root locus.

$$(R_p - R_z)180° = (1+2m)180°$$

Where R_p = number of poles to the right of the search point on the real axis and R_z = number of zeros.

C) End points of the locus:

The starting points of the root-locus (i.e. for k = 0) are the open-loop poles and the end points (for k = ∞) are the open-loop zeros (∞ is an equivalent zero).

Asymptotes of the root loci (behavior as $s \to \infty$):

$$\theta = \frac{(1 + 2m)180°}{[\text{number of poles of } G(s)H(s)] - [\text{number of zeros of } G(s)H(s)]}$$

= This is the angle of straight lines or asymptotes.

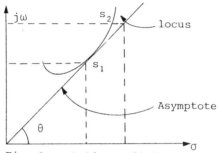

Fig. Asymptotic condition

The behavior of the root-locus near ∞ are important because when number of poles of $G(s)H(s) \neq$ number of zeros of $G(s)H(s)$, then 2 |number of poles $G(s)H(s)$ − number of zeros of $G(s)H(s)$| will tend to infinity in the s-plane.

Intersection of the asymptotes (centroid):

$$\sigma_0 = \text{Real axis interception of the asymptotes} = \frac{\sum_{c=1}^{v} \text{Re}(P_c) - \sum_{m=1}^{w} \text{Re}(Z_m)}{v - w}$$

since the proportions of the locus away from the asymptotes and near the axes are important. σ_0 is also the centroid of the pole zero plot where

$$A(s) = \frac{\prod_{c=1}^{v} (s - p_c)}{\prod_{m=1}^{w} (s - z_m)} = -k$$

Note that there are 2|number of poles of $G(s)H(s) = v$ - number of zeros ϕ of $G(s)H(s) = w$|.

Asymptotes whose intersection lies on the real axis.

Also note that the point of intersection is always on the real axis.

Real axis break-away points (saddle points):

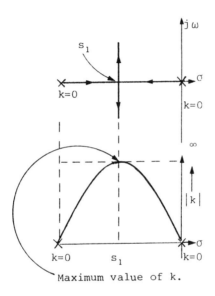

The break-away point: Since k starts with a value of zero at the poles and increases in value, as the locus moves away from the poles, there is a point somewhere in between where the k's for the two branches simultaneously reach a maximum value. This is the break-away point.

Break-away points on the root-loci of an equation correspond to multiple-order roots of the equation.

A root locus can have more than one break-away point and these need not always be on the real axis; however the break-away points may be real or complex conjugate pairs.

The break-in point: This is the value of σ for which $|k|$ is a minimum between 2 zeros. This is shown in the diagram.

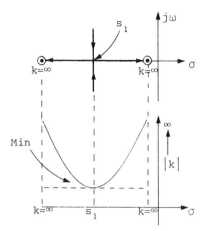

Break-in points

Break-away points: These can be determined by taking the derivative equating to zero and then determining its roots. The root that occurs between the poles (or the zeros) is the break-away (or break-in) point.

Example: $G(s)H(s) = k/s\,(s+1)(s+2)$

so $W(s) = s(s+1)(s+2) = -k = s^3 + 3s^2 + 2s$

$$\frac{dW(s)}{ds} = 3s^2 + 6s + 2 = 0$$

Hence, the roots are:

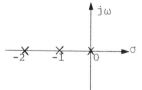

Fig. Pole-zero plot.

$s_{1,2} = -1 \pm 0.6$ i.e. -0.4 and -1.57

The break-away point for $k > 0$ lie between $s = 0$ and $s = -1$ in order to satisfy the angle condition, $s_1 = -0.4$. The other point, $s_2 = -1.57$ is the break-in point for $k < 0$.

$$-k = (-0.4)^3 + 3(-0.4)^2 + 2(-0.4)$$

so, $k = 0.38$

(This is the value of k at the break-away point for $k > 0$.)

The necessary angular condition near the break-away point: A break-away point is determined as follows:

Step 1: A check point is selected between two poles on the real axis at a distance d; then the angular contribution due to all poles and zeros which are on the real axis is $\frac{d}{L_i}$, where L_i is the distance between the check point and the pole or zero.

Step 2: The angular contribution towards this check point due to complex poles or zeros is

$$\Delta \phi = 2b \delta /b^2 + a^2 \quad \text{as shown}$$

Step 3: All the angular contributions with proper signs are summed to zero. The point which satisfies this zero condition is the break-away point.

Complex pole (or zero): Angle of departure

For $k > 0$, the angle of departure from a complex pole is equal to $180°(1 + 2m)$ minus the sum of the angles from the other poles plus the sum of the angles from the zeros. Any of these angles may be +ve or -ve. For $k < 0$, the departure angle is $180°$ from that obtained for $k > 0$.

Imaginary-axis intercepting point:

The points where the complete root locus intersects the imaginary axis of the s-plane, and the corresponding values of k can be determined by means of the Routh-Hurwitz criterion:

Example: $s^3 + as^2 + bs + kd = 0$

The closed-loop characteristic equation.

A Routhian array is formed

$$\begin{array}{c|cc} s^3 & 1 & b \\ s^2 & a & \\ s^1 & (ab-kd)a & kd \\ s^0 & kd & \end{array}$$

The Routhian array formed from the closed-loop characteristic equation.

Undamped oscillation occurs if the s^1 row is zero. Thus, the auxiliary equation obtained from the s^2 row is:

$$as^2 + kd = 0 \quad \text{and its roots} \quad s_1 = +j\sqrt{\frac{kd}{a}} \quad \text{and} \quad s_2 = -j\sqrt{\frac{kd}{a}}$$

i.e. $\quad k = \dfrac{ab}{d}$

Root locus branches (non-intersection and intersection):

Properties

A) Any point on the root locus satisfies the angle condition. There are no root locus intersections at a point on the root locus if $\dfrac{dW(s)}{ds} \neq 0$ at this point. In this case there is only one branch of the root locus through the point.

B) A point on the root locus will have branches through it (i.e. it is an intersection point) if the derivatives of $W(s)$ vanish at this point. Thus if the first y-1 derivatives of $W(s)$ vanish at a given point on the root locus, there will be y branches approaching and y branches leaving this point. The angle between 2 approaching branches is

$$\lambda_y = \pm \frac{360°}{y}$$

while the angle between 2 branches (one leaving and the other approaching the same point) is

$$\phi_y = \pm \frac{180°}{y}$$

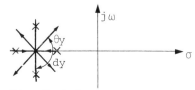

Fig. Root locus.

Conservation of the sum of the system of roots:

Grant's rule states that the sum of the closed-loop roots is equal to the sum of the open-loop poles. This is applicable when the open-loop transfer function is such that $v - z \geq W$.

Determination of roots on the root locus:

Application of Grant's rule

Grant's rule which is stated as follows, is used to find one real or two complex roots of the system provided the dominant roots of the characteristic equations is known.

$$\sum_{k=1}^{v} p_k = \sum_{l=1}^{v} r_l$$

7.3 FREQUENCY RESPONSE

DEFINITION

The frequency response gives the ratio of the phasor output to the phasor input for any inputs over a range of frequencies. The combined plots of the magnitude (M) and the angle (α) of $\frac{C(j\omega)}{R(j\omega)}$ versus the angular frequency is the frequency response of a control system. How to obtain the frequency response:

a) To determine the frequency response, the closed-loop control ratio should be known.

b) Then the control ratio is expressed as a function of frequency by substituting $s = j\omega$ as follows.

$$\frac{C(j\omega)}{R(j\omega)} = \frac{k(j\omega + a)}{(j\omega + b - cj)(j\omega + b + c)(j\omega + d - ej)(j\omega + d + ej)}$$

Please note that this is the most generalized form.

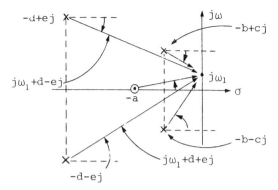

Fig. Frequency response from the plot.

c) For any frequency ω_1, a point ω_1 is chosen on the $j\omega$ axis, and the directed lines are drawn from all poles and zeros to this point. Then the lengths of the directed lines and angles these lines make with horizontal lines are determined as shown in the figure.

d) When the magnitudes and angles for each term (i.e. for each of the directed lines) of the $\frac{C(j\omega)}{R(j\omega)}$ equation are obtained, the value of $C(j\omega)/R(j\omega)$ is obtained for that particular frequency ω_1. Note that the clockwise angles are -ve and the anticlockwise angles are +ve.

e) This procedure is repeated for sufficient numbers of angular frequencies and a smooth curve is drawn as shown below.

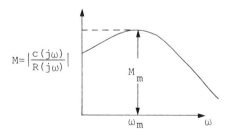

CHAPTER 8

SPECIAL POLE-ZERO TOPICS: DOMINANT POLES AND THE PARTITION METHOD

8.1 TRANSIENT RESPONSE: DOMINANT COMPLEX POLES

The conditions which are necessary for the time response to be nominated by only one pair of complex poles, require a pole-zero pattern with the following characteristics:

A) All other poles in the pole-zero diagram must be far to the left of the dominant poles, so the transient response due to these poles are small in amplitude.

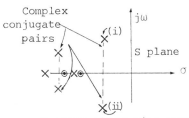

Pattern whose time response is dominated by complex poles.

B) In the pole-zero diagram, any pole which is not far away from the dominant complex poles must be near a zero, so that the magnitude of the transient response is small; since its effect will be modified by the zero. Note: The dominant poles are drawn darker.

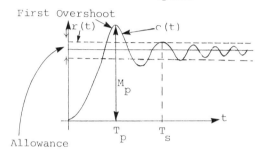

The Time Response

FIGURES OF MERIT

A) M_p (peak overshoot): The amplitude of the first overshoot.

B) t_p (peak time): The time to reach the peak overshoot from the initial starting time.

C) t_s (setting time): The time for the response to first reach and thereafter remain within the allowable limit (usually 2% of the final value).

D) n: The number of oscillations in the response up to the setting time. Not that there are two oscillations in a complete cycle.

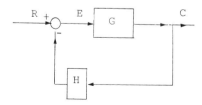

Non-Unity Feedback System

Figures of merit for the non-unity feedback system:

$$G(s)H(s) = \frac{k_G k_H \prod_{m=1}^{\omega}(s - z_m)}{\prod_{i=1}^{v}(s - p_i)}$$

where product $k_G k_H = k$ is the static loop sensitivity.

$$\frac{C(s)}{R(s)} = \frac{k_G \prod_{m=1}^{n}(s - z_n)}{\prod_{i=1}^{v}(s - p_i)}$$

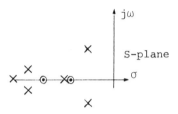

The desired Pole-Zero pattern

Here k_G is not k unless the system has unity feedback.

$$= \frac{A_0}{s} + \frac{A_1}{s - p_1} + \ldots$$

$$c(t) = \begin{array}{l}\text{Time solution if the}\\ \text{system has a dominant}\\ \text{pole } p_1 = \sigma + j\omega_d\end{array} = \frac{P(o)}{Q(o)} + 2\left|\frac{k_G \prod_{m=1}^{n}(p_1 - z_m)}{p_1 \prod_{i=2}^{v}(p_1 - p_i)}\right|$$

$$e^{\sigma t} \cdot \cos[\omega_d t + \underline{/P(p_1)} - \underline{/p_1} - \underline{/Q'(p_1)}]$$

$$+ \sum_{k=3}^{v} \frac{P(p_k) e^{p_k t}}{p_k Q'(p_k)}$$

where
$$Q'(p_k) = \left.\frac{Q(s)}{s-p_k}\right]_{s=p_k}$$

$$T_p = \frac{1}{\omega_d}\left\{\frac{\pi}{2} - \begin{bmatrix}\text{sum of angles} \\ \text{from zeros of } \frac{C(s)}{R(s)} \\ \text{to } P_1, \text{ the dominant} \\ \text{pole}\end{bmatrix} + \begin{bmatrix}\text{sum of angles from} \\ \text{all other poles of } \frac{C(s)}{R(s)} \\ \text{to dominant pole } p_1 \\ \text{(including conjugate} \\ \text{pole)}\end{bmatrix}\right\}$$

$$M_p = \text{The peak overshoot} = \underbrace{\frac{P(o)}{Q(o)}}_{\substack{\text{The} \\ \text{final} \\ \text{value}}} + \underbrace{\frac{2\omega_d}{\omega_n^2}\left|\frac{k_G \prod_{m=1}^{\omega^1}(p_1 - z_m)}{\prod_{i=2}^{v}(p_1 - p_i)}\right|e^{\sigma T_p}}_{\text{The overshoot } M_0}$$

$$\boxed{\begin{aligned}t_s &= \frac{m}{|\sigma|} = \frac{m}{\zeta\omega_n} = \begin{array}{l}\text{Four time constants} \\ \text{for 2\% error, i.e., } m = 4\end{array} \\ n &= \frac{\text{Settling time}}{\text{Period}} = \frac{t_s}{2\pi/\omega_d} = \frac{2\omega_d}{\pi|\sigma|} = \frac{2}{\pi}\frac{\sqrt{1-\zeta^2}}{\zeta}\end{aligned}}$$

Additional Significant Poles:

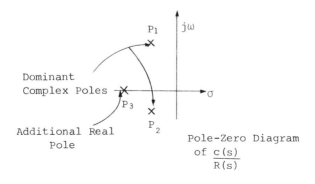

Pole-Zero Diagram of $\frac{c(s)}{R(s)}$

$$c(t) = 1 + 2|A_1|e^{-\zeta\omega_n t}\sin(\omega_n\sqrt{1-\zeta^2}\,t + \phi) + A_0 e^{P_0 t}$$

The time response to a unit step input for a system,

$$\frac{C(s)}{R(s)} = \frac{k}{(s^2 + 2\zeta\omega_n s + \omega_n^2)/(s + P_0)}$$

The effects of an additional real pole can be given as follows:

Time Responses As A Function of Real-Pole Location

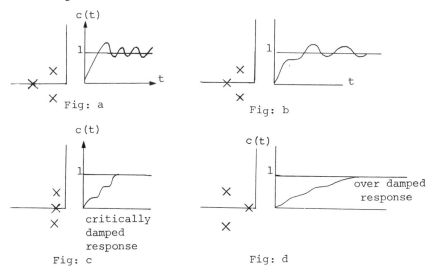

A) Peak overshoot M_p is reduced.

B) The setting time t_s is increased because $A_0 e^{P_0 t}$ is the transient term due to P_0 and A_0 is negative.

C) $|A_0|$ depends on the relative location of P_0 with respect to dominant pole of the distance between P_0 and complex poles is large, then A_0 is small.

Effects of additional real pole and zero:

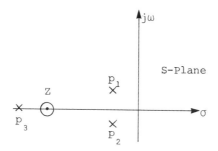

Fig. Pole-zero plot.

A) The complete time response to a unit-step function has the same form as in the case of an additional real pole.

B) Sign of A_o depends on the relative locations of the real pole and the real zero. A_o is negative if the zero is to the left of p_o and is positive if the zero is to the right of p_o. A_o is proportional to the distance from p_o to z.

C) If the zero is close to the pole, A_o is small, then the contribution of this transient term is small.

8.2 POLE-ZERO DIAGRAM AND FREQUENCY AND TIME RESPONSE

CHARACTERISTICS

A) Fig. a:
   a) Frequency response curve has a single peak M_m and $1.0 < M < M_m$ in the frequency range $0 < \omega < 1.0$.
   b) Time response: The first maxima of $c(t)$ due to the oscillatory term is greater than $c(t)_{ss}$ and the $c(t)$ response after this maxima oscillates around the value $c(t)_{ss}$.

Relation: An illustration.

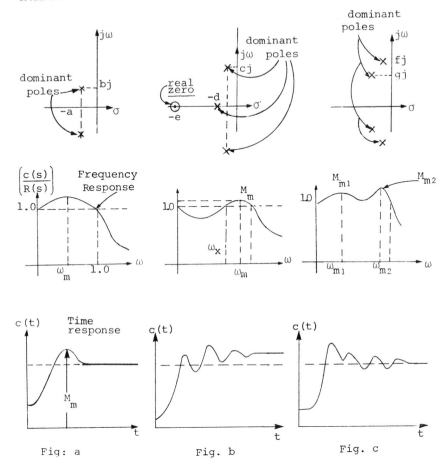

Fig:-Comparison of frequency and time responses

B) Fig. b: Time response: The first maximum of c(t) due to oscillatory term is less than $c(t)_{ss}$.

C) Fig. c: Time response: The first maxima of c(t) in the oscillation is greater than $c(t)_{ss}$, and the oscillatory portion of c(t) does not oscillate about a value of $c(t)_{ss}$.

The system's time-response can be predicted to a

greater extend from the shape of the frequency-response plot.

Table 8.1

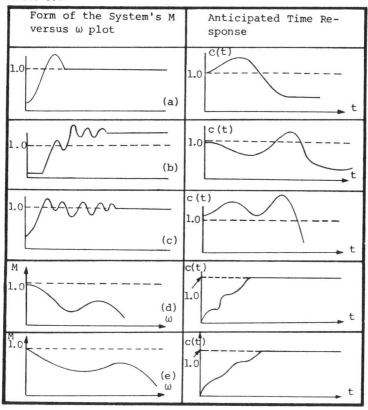

Correlation between Frequency and Time Response

8.3 FACTORING OF POLYNOMIALS USING ROOT-LOCUS

Partition method: The application of the root-locus method for factoring polynomials is called the partition method.

An example:

$$s^3 + es^2 + fs + g = 0$$

where e, f and g are constants.

A) This can be partitioned at s^3, s^2 and s.

B) Partition at s^3:

$$s^3 = -(es^2 + fs + g)$$

$$-1 = \frac{e(s^2 + \frac{f}{e}s + \frac{g}{e})}{s^3} = I_1(s)$$

$$= \frac{e(s + \beta)(s + \gamma)}{s^3}$$

C) Partition at s^2:

$$s^3 + es^2 = -(fs + g)$$

$$-1 = \frac{f(s + \frac{g}{f})}{s^2(s + e)} = I_2(s)$$

D) $s^3 + es^2 + fs = -g$

$$-1 = \frac{g}{s(s^2 + es + f)} = \frac{g}{s(s + \alpha)(s + \sigma)} = I_3(s)$$

E) Each resulting equation after partitioning has a form $I(s) = -1$. It should satisfy the angle and magnitude conditions. Since the resulting equation looks like $G(s)H(s) = -1$, the root-locus method can be utilized to determine the roots of a polynomial.

F) For a third-degree polynomial, partition at s^2 is preferred.

HANDBOOK of MATHEMATICAL, SCIENTIFIC, and ENGINEERING FORMULAS, TABLES, FUNCTIONS, GRAPHS, TRANSFORMS

A particularly useful reference for those in math, science, engineering and other technical fields. Includes the most-often used formulas, tables, transforms, functions, and graphs which are needed as tools in solving problems. The entire field of special functions is also covered. A large amount of scientific data which is often of interest to scientists and engineers has been included.

Available at your local bookstore or order directly from us by sending in coupon below.

RESEARCH and EDUCATION ASSOCIATION
61 Ethel Road West • Piscataway, N.J. 08854
Phone: (201) 819-8880

VISA MasterCard

Please check one box:
☐ Check enclosed
☐ Visa
☐ MasterCard

Charge Card Number ☐☐☐☐☐☐☐☐☐☐☐☐☐☐☐

Expiration Date (Mo./Yr.) _____

Please ship the "Math Handbook" @ $21.85 plus $2.00 for shipping.

Name...
Address..
City......................................State.................

THE PROBLEM SOLVERS

The "PROBLEM SOLVERS" are comprehensive supplemental textbooks designed to save time in finding solutions to problems. Each "PROBLEM SOLVER" is the first of its kind ever produced in its field. It is the product of a massive effort to illustrate almost any imaginable problem in exceptional depth, detail, and clarity. Each problem is worked out in detail with step-by-step solution, and the problems are arranged in order of complexity from elementary to advanced. Each book is fully indexed for locating problems rapidly.

ADVANCED CALCULUS
ALGEBRA & TRIGONOMETRY
AUTOMATIC CONTROL SYSTEMS/ROBOTICS
BIOLOGY
BUSINESS, ACCOUNTING, & FINANCE
CALCULUS
CHEMISTRY
COMPLEX VARIABLES
COMPUTER SCIENCE
DIFFERENTIAL EQUATIONS
ECONOMICS
ELECTRICAL MACHINES
ELECTRIC CIRCUITS
ELECTROMAGNETICS
ELECTRONIC COMMUNICATIONS
ELECTRONICS
FINITE and DISCRETE MATH
FLUID MECHANICS/DYNAMICS
GENETICS

GEOMETRY: PLANE • SOLID • ANALYTIC
HEAT TRANSFER
LINEAR ALGEBRA
MACHINE DESIGN
MECHANICS: STATICS • DYNAMICS
NUMERICAL ANALYSIS
OPERATIONS RESEARCH
OPTICS
ORGANIC CHEMISTRY
PHYSICAL CHEMISTRY
PHYSICS
PRE-CALCULUS
PSYCHOLOGY
STATISTICS
STRENGTH OF MATERIALS & MECHANICS OF SOLIDS
TECHNICAL DESIGN GRAPHICS
THERMODYNAMICS
TRANSPORT PHENOMENA: MOMENTUM • ENERGY • MASS
VECTOR ANALYSIS

If you would like more information about any of these books, complete the coupon below and return it to us or go to your local bookstore.

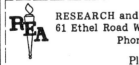

RESEARCH and EDUCATION ASSOCIATION
61 Ethel Road West • Piscataway, N.J. 08854
Phone: (201) 819-8880

Please send me more information about your Problem Solver Books.

Name ....

Address ....

City .... State ....